TERRAFORMING

Liverpool Science Fiction Texts and Studies, 55

Liverpool Science Fiction Texts and Studies

Recent titles in the series

30. Mike Ashley *Transformations: The Story of the Science-Fiction Magazine from 1950–1970*
31. Joanna Russ *The Country You Have Never Seen: Essays and Reviews*
32. Robert Philmus *Visions and Revisions: (Re)constructing Science Fiction*
33. Gene Wolfe (edited and introduced by Peter Wright) *Shadows of the New Sun: Wolfe on Writing/Writers on Wolfe*
34. Mike Ashley *Gateways to Forever: The Story of the Science-Fiction Magazine from 1970–1980*
35. Patricia Kerslake *Science Fiction and Empire*
36. Keith Williams *H. G. Wells, Modernity and the Movies*
37. Wendy Gay Pearson, Veronica Hollinger and Joan Gordon (eds.) *Queer Universes: Sexualities and Science Fiction*
38. John Wyndham (eds. David Ketterer and Andy Sawyer) *Plan for Chaos*
39. Sherryl Vint *Animal Alterity: Science Fiction and the Question of the Animal*
40. Paul Williams *Race, Ethnicity and Nuclear War: Representations of Nuclear Weapons and Post-Apocalyptic Worlds*
41. Sara Wasson and Emily Alder, *Gothic Science Fiction 1980–2010*
42. David Seed (ed.), *Future Wars: The Anticipations and the Fears*
43. Andrew M. Butler, *Solar Flares: Science Fiction in the 1970s*
44. Andrew Milner, *Locating Science Fiction*
45. Joshua Raulerson, *Singularities*
46. *Stanislaw Lem: Selected Letters to Michael Kandel* (edited, translated and with an introduction by Peter Swirski)
47. Sonja Fritzsche, *The Liverpool Companion to World Science Fiction Film*
48. Jack Fennel: *Irish Science Fiction*
49. Peter Swirski and Waclaw M. Osadnik: *Lemography: Stanislaw Lem in the Eyes of the World*
50. Gavin Parkinson (ed.), *Surrealism, Science Fiction and Comics*
51. Peter Swirski, *Stanislaw Lem: Philosopher of the Future*
52. J. P. Telotte and Gerald Duchovnay, *Science Fiction Double Feature: The Science Fiction Film as Cult Text*
53. Tom Shippey, *Hard Reading: Learning from Science Fiction*
54. Mike Ashley, *Science Fiction Rebels: The Story of the Science-Fiction Magazines from 1981 to 1990*

TERRAFORMING

Ecopolitical Transformations and Environmentalism in Science Fiction

CHRIS PAK

LIVERPOOL UNIVERSITY PRESS

First published 2016 by
Liverpool University Press
4 Cambridge Street
Liverpool
L69 7ZU

British Library Cataloguing-in-Publication data
A British Library CIP record is available

ISBN 978-1-78138-284-4 cased

Typeset by Carnegie Book Production, Lancaster
Printed and bound by CPI Group (UK) Ltd, Croydon CR0 4YY

For Irene, Alex, and Dennis

Contents

Acknowledgements

This book grew out of research conducted at the Department of English at the University of Liverpool. I owe a special thanks to Andy Sawyer for his mentorship, and for introducing me to the academic study of science fiction. The staff at the Special Collections and Archives at the University of Liverpool's Sydney Jones Library also deserve a thank you for their assistance in providing access to the research materials available at the Science Fiction Foundation Collection that made this book possible. I had the pleasure of exchanging emails with Frederick Turner, who generously answered many questions about his work and about terraforming and landscape restoration, and who kindly agreed to be interviewed for the *SFRA Review*. I also want to thank Siobhan Chapman, David Seed, and Mark Bould for their criticism of the PhD thesis that eventually became this book, and Simon Hailwood for introducing me to the domain of environmental philosophy.

I am indebted to my colleagues, both academic and administrative, for providing an environment that helped shape my intellectual life throughout the nine years that I spent at Liverpool. I have received much encouragement and stimulation from many individuals, not least Katherine Buse, Molly Cobb, Barry Dainton (at the Department of Philosophy), Nick Davis, Leimar Garcia-Siino, David McWilliam, Glyn Morgan, and Michelle Yost. The Science Fiction Foundation, the Science Fiction Research Association, and the Association for the Study of Literature and the Environment – UK and Ireland were instrumental in shaping my thinking about the intersections between science fiction and ecology. I would like to thank these organisations for their support, and in particular Thierry Bardini, Ritch Calvin, Paweł Frelik, Greg Garrard, Joan Haran, Ursula K. Heise, Larisa Mikhaylova, T.S. Miller, John Parnham, Adeline-Johns Putra, Louise Squire, Sherryl Vint, and Liz Williams. Thanks are also due to Mariano Martín Rodríguez, and to

Cathy Rees, Chris Williams, and Phil Franey, whose support throughout my time at the university is much appreciated.

Elements of the Introduction appear in the journal publication 'Environmental Philosophy and Space in Literature: Landscaping and the Chronotope,' *Brief: Online Journal of Snippets – American Literature*, 1 (2011), 5–6 <http://www.brief.umcs.eu/american-literature/brief-1_american-literature.pdf> (accessed March 2016).

Elements of the first chapter appear in the journal publications '"A Creature Alive but Tranced and Obscurely Yearning to Wake": Gaian Anticipations and Terraforming in the European Science Fiction of H.G. Wells, Olaf Stapledon and Stanisław Lem,' *Revista Hélice: Reflexiones Críticas Sobre Ficción Especulativa*, 15 (2012a), 12–19; '"A Fantastic Reflex of Itself, An Echo, A Symbol, A Myth, A Crazy Dream": Terraforming as Landscaping Nature's Otherness in H.G. Wells's *The Shape of Things to Come* and Olaf Stapledon's *Last and First Men* and *Star Maker*,' *Foundation: The International Review of Science Fiction*, 111.40 (2012b), 14–31; and 'Terraforming and Proto-Gaian Narratives in American Pulp Sf of the 1930s–1940s,' *Eaton Journal of Archival Research*, 1 (2013), 38–55.

Elements of the fifth chapter appear in the journal publications 'Ecocriticism and Terraforming: Building Critical Spaces,' *Forum: University of Edinburgh Postgraduate Journal of Culture & the Arts*, 10 (2010) <http://www.forumjournal.org/site/issue/10/chris-pak> (accessed March 2016); and '"All Energy Is Borrowed" – Terraforming: A Master Motif for Physical and Cultural Re(Up)cycling in Kim Stanley Robinson's *Mars* Trilogy,' *Green Letters*, 18 (2014a), 1–13 <http://www.tandfonline.com/doi/abs/10.1080/14688417.2014.890527> (accessed March 2016).

Introduction

Terraforming: Engineering Imaginary Environments

Science-fictional (sf) stories of planetary adaptation – terraforming – construct imaginative spaces to explore society's orientation to ecological, environmental, and geopolitical issues and concerns. Terraforming involves processes aimed at adapting the environmental parameters of alien planets for habitation by Earthbound life, and it includes methods for modifying a planet's climate, atmosphere, topology, and ecology. Combining the Latin *terra* for 'earth' or 'land' with the gerund 'forming,' the term refers to '[t]he process of transforming a planet into one sufficiently similar to the earth to support terrestrial life' and is chiefly associated with sf discourse ('Terraforming, n.,' 2015). While the *Online OED* credits the first use of the term to Jack Williamson (writing as Will Stewart) in 1949, Jeffrey Prucher traces the verb 'terraform' to Williamson's 1942 short story 'Collision Orbit' (2007, 235). To a primary definition similar to the *Online OED*'s, Prucher adds two others: 'to modify a world's environment so that it can support life that evolved on a planet other than the Earth' (dated to 1969) and 'to modify the Earth's environment' (dated to 1997) (235).

These definitions encompass three modes of terraforming. The first designates the human colonisation of space where alien planets are shaped in the image of Earth. The second involves an alien colonisation of space and the alteration of planets to resemble the aliens' homeworlds. The third, the alteration of Earth's landscape, is at first glance puzzling when paired with the *Online OED*'s primary definition: what does it mean to alter Earth to make it more closely resemble itself? Martyn Fogg helpfully defines terraforming alien planets and terraforming Earth, or 'geoengineering,' as two subsets of 'planetary engineering,' arguing that 'phrases such as "terraforming the Earth" have a ring of nonsense about them – how does one make the Earth more like itself?' (1995, 90). He explains that '[g]eoengineering is planetary engineering applied specifically to the Earth. It includes only those macroengineering concepts that deal with the alteration of some global parameter, such as the greenhouse effect, atmospheric composition, insolation or impact flux' (90).

Prucher's primary definition offers a clue to this conundrum: 'to modify a world's environment so that it can support Earth life-forms, especially humans' (2007, 235). When considered against the definition of the adjectival form of the term, 'terraformed,' '(of a world) having been modified to support life-forms alien to it,' avenues for reframing orientations and perspectives towards the habitation of Earth are opened (235). These definitions encode a conception of humanity as fundamentally alien to Earth. Modification through agricultural technologies could be considered early instances of an impulse to shape the planet for human-centred purposes, culminating in images and narratives that feed into real-world motivations to terraform other planets.

The date of Prucher's third definition reflects a shift in awareness of humankind's ability to alter planets through climate change and other global effects. The link between terraforming, geoengineering and climate change has been articulated by many scientists and commentators on contemporary environmental issues. Christopher McKay suggests that 'it is becoming increasingly clear that humanity is already engaged in both deliberate and inadvertent global modifications of at least one planet – Earth' (1982, 309), while Michael Dumiak explains that '[t]erraforming Mars is basically a radical application of human-induced climate change' (2007, 62). This sense of terraforming as an extension of anthropogenic climate change illustrates a connection between climate change and geoengineering, and by a further conceptual extension geoengineering and terraforming.

Terraforming as a narrative, a motif, and a concept exemplifies the feedback between sf, science, and wider popular culture. The term was adopted by scientific discourse, before feeding into sf through borrowings from scientific speculation about terraforming. In his scientific survey of planetary adaptation, Fogg explains that the technical study of terraforming involves environmental, social, political, legal and ethical complexities that impact on real-world considerations of planetary adaptation (1995, 24). Although he acknowledges the root of the theme in sf and includes a short discussion of such literature, Fogg's focus as a scientist remains largely on the technical possibilities for terraforming. While the dialogue between sf and science is central to the theme, it has also gained traction in both environmental and wider popular spheres.

Given the impossibility of human evolution that would be swift enough to allow the safe habitation of other planets, the colonisation of other worlds depends on a series of technological adaptations. In the same year that Williamson coined 'terraforming,' James Blish coined 'pantrope,' loosely translated as 'changing everything,' to refer to the genetic modification of humans for habitation of alien environments (2001 [1956], 8). Pantropy has since been expanded to include biotechnologies alternative to genetic engineering, such as cyborgisation. Terraforming involves the adaptation of space via industrial methods, through a series of ecologically informed adaptations, or through a combination of both. Terraforming and pantropy can also be combined, such as when humans are genetically engineered to cope better with the climate of the terraformed yet still alien Mars in Kim Stanley Robinson's acclaimed *Mars* trilogy (1996c [1992]; 1996b [1993]; 1996a).

Prucher's three definitions are useful guides to tracking how terraforming is imagined in 1942 as part of a human colonisation of space, before it becomes established by 1969 as inclusive of any civilisation and, implicitly, the alternative worlds that act as models guiding alien planetary adaptation. By 1997 the imaginative spaces offered to environmental speculation by sf narratives of terraforming are reconnected to Earth, another development that has influenced wider scientific and philosophic discourse amidst contemporary anxieties about environmental change. These changes to the context that terraforming is understood in and the way they emphasise different aspects of an ecopolitical intersection suggest that sf has continually shifted focus in response to new conceptions of human relationships to physical, value-laden spaces. Terraforming provides a clear example of the dialogue between sf, science and environmentalism that makes it especially significant for ecocritical examination. This book is not about the scientific and technical aspects of terraforming but about terraforming as a motif and a narrative that engages with ecology and environmentalism, socio-politics and ethics. Pantropic themes are not analysed in equal depth for reasons of space, but at times it will be important to consider how pantropy, as a supplement or alternative to terraforming, foregrounds concerns that underlie decisions to physically adapt other worlds.

Shaping Earth and the Solar System

> Making your world more habitable began on the Earth itself, with the first dancing fire that warmed its builder's cave. (Reed, 2001, 199)

The significance of geoengineering and its overlap with terraforming makes it important to keep in focus their correspondences and distinctions. The historians John McNeill and J. Donald Hughes discuss the effects on the environment of such projects as the Boulder Dam on the Colorado River (renamed the Hoover Dam) and the Aswan High Dam across the Nile, along with other related endeavours such as urbanisation and rapid technological change. The relationship between terraforming and geoengineering is implied by the similarity of language used by advocates of large-scale engineering projects to justify these developments. As McNeill and Hughes argue, projects such as dam construction are often motivated by political agendas that sometimes exceed strictly economic concerns (McNeill, 2001, 157–82; Hughes, 2009, 175–81).

In M. Vassiliev and S. Gouschev's speculative account of the benefits that Soviet science might bring to Russia in the future, geoengineering themes that relate directly to the adaptation of landscapes are anticipated with delight. In the chapter 'The Dawning Age of Plenty,' several 'glimpses' of the future are described, including 'the enrichment of the soil, the promotion of rain, and the conversion of salt lakes and inland seas into fresh-water ones' so that desert environments can be adapted into fertile agricultural land (1961, 94). In 'The Creators of Nature,' the global distribution of various natural resources is criticised in order to buttress an argument advocating the transformation of 'our communal house, the earth,' to the needs and tastes of humankind (185). While such transformations of nature have already occurred in Russia, it is predicted that, 'as man's power increases, this transforming activity will develop even further' (186). 'In the Lunar City' takes the logic of the technological transformation of nature into space, recounting as it does the construction of dome-like glass cities on the Moon (205–08).

Adrian Berry focuses particular attention on terraforming and astrophysical engineering in his popular scientific work, *The Next Ten Thousand Years*, citing examples of such projects in sf by Olaf Stapledon and Poul Anderson (1976, 93, 91). Berry critiques the apocalyptic strand of environmentalist discourse, which he sees culminating in the publication of *The Limits to Growth* in 1972, the 'Rousseau-like dreams [...] of a "return to nature," and the desirability of living without technical aids like the "noble savage"' (187). Sir Francis Bacon's notion of progress and the publication of his scientific utopia *The New Atlantis* in 1627 are significant in Berry's view because '[i]t was perceived for the first time that humanity might have a hidden purpose, and might be able to execute a long-term plan whose nature had been hitherto concealed' (23). Berry argues that terraforming and the colonisation of space are precisely such activities and concludes, in contrast to warnings of economic and environmental decline, that '[t]he Baconian scheme can be delayed, but it cannot be stopped' (189).

Sf as Environmental Literature

Environmentalism is a broad term involving several areas of discourse. In the nineteenth and early twentieth centuries, environmentalism had been preoccupied with issues of conservation and preservation. Henry David Thoreau's *Walden*, published in 1854, greatly influenced the environmental movement, as did Ralph Waldo Emerson's American transcendental poetry. The Sierra Club, founded in 1892 by John Muir, did much to establish conservation in America. Samuel Butler's 1872 *Erewhon* portrays a future pastoral utopia where machines have been banned because of the perceived threat that their evolution poses to humankind. Richard Jefferies's 1885 *After London* is an early work of scientific romance set in a post-apocalyptic England divided into feudal counties. It possesses a strong environmental orientation and contrasted the new pastoral landscapes of this future with a submerged and contaminated region where London used to stand. In W.H. Hudson's 1887 *A Crystal Age*, a matriarchal society of independent family households live in a radically depopulated far future. The narrative follows the struggles of a protagonist, brought by an accident into this vividly depicted pastoral landscape, who fails to adapt his contemporary values to the expectations of his adoptive household. Responding to Edward Bellamy's 1888 *Looking Backward, 2000–1887*, William Morris's 1890 socialist utopia, *News From Nowhere*, depicts a deurbanised pastoral England visited – as with Hudson's and Bellamy's works – by a visitor contemporary to the narrative's publication. Since the 1960s, ecology has played a larger role in informing environmental awareness, shifting this early investment in preservation and conservation in new directions.

Susan Stratton suggests that sf owes more to environmentalism as a separate collection of social, political and economically focused movements or projects of protest, activism, reformation and deconstruction than it does to any relatively unified notion of environmentalism (2000, 2–6). *Silent Spring* (2002), Rachel Carson's foundational exposé of the far-reaching chemical pollution of the ecosystem, is often cited as the popularising text that brought environmental issues to the forefront of an international awareness. Paul Ehrlich's *The Population Bomb* (1971) and the first images of Earth taken by the US Weather Bureau's TRIOS satellite, launched in the 1960s, also highlighted concerns that were popularised by the first Earth Day in 1970 (Daly and Frodeman, 2008, 136). Environmental philosophy grew out of, and to some extent alongside, a burgeoning 'activist' strand supported by environmentally focused philosophies. Debates in environmental philosophy tend to centre primarily on issues of ethics and value, although aesthetics, theology and ecofeminism are also dominant areas of philosophical enquiry.

Both environmental philosophy and ecocriticism developed in response to the growth of environmental awareness, but only became established as academic disciplines in the 1980s. Ecocriticism is a form of literary criticism focused on the contribution to environmentalism of literary texts, and it involves literary-aesthetic as well as philosophic examination of the relationships between humanity and the environment. The term 'ecocriticism' was coined by William Rueckert in his 1978 article 'Literature and Ecology: An Experiment in Ecocriticism,' since collected in *The Ecocriticism Reader* (1996, 105–23). Nevertheless, several works published before 1978 anticipated the concerns of ecocriticism: Raymond Williams's *The Country and the City* (2011 [1973]) includes a chapter on sf and Annette Kolodny's retrospectively ecofeminist *The Lay of the Land* (1975) examines the gendered discourse involved in the colonisation of America.

In *Trillion Year Spree,* Brian Aldiss and David Wingrove define sf as *'the search for a definition of mankind and his status in the universe which will stand in our advanced but confused state of knowledge (science)'* (2001, 4). By characterising sf as a 'search for a definition,' Aldiss claims that sf is a literature of epistemology, an exploration of how science and technology force a re-evaluation of humankind's place in relation to its environment and the cosmos. Aldiss argues for sf's potential value as environmental literature by claiming that '[t]he greatest successes of science fiction are those which deal with man in relation to his changing surroundings and abilities: what might loosely be called *environmental fiction'* (8). His emphasis on a loosely defined 'environmental fiction' highlights the wide range of issues that inform an environmental awareness while implicitly acknowledging the multiple ways sf explores two major themes that are essential to the mode: our relationship to the environment, and the way our abilities – our technologies – allow alteration of both the environment and the range of environments made available to us.

Noel Gough argues that sf's focus on the external world and on our interaction with it is a result of sf's object orientation. Like Aldiss, he claims that 'this attention to externalities may mark SF as an environmental literature par excellence' (1998, 411). Patrick D. Murphy concurs when he calls sf a nature-oriented literature. Like Aldiss and Gough, he argues that this is because it 'directs reader attention toward the natural world and human interaction with other aspects of nature within that world,' but he also points out that it 'makes specific environmental issues part of the plots and themes of various works' (Murphy, 2001,

263). Terraforming is a significant instance of such environmental issues because the motif flexibly accommodates a range of environmental events, thus opening up a potentially vast field for environmental philosophical speculation.

Sf, as the example of terraforming illustrates, is a mode that allows us to explore the status and the consequences of various forms of relationship to space. While consideration of these issues in terraforming stories is usually focused on our attitudes to planets other than Earth, such stories allow us to examine and evaluate our historical relationship to our home planet and to postulate alternatives to current practices. The 'terra' in 'terraforming' always refers us to its paradigmatic example, especially when used in sf discourse. The extremes of spatial and temporal scale explored in terraforming stories allow us to imaginatively re-situate our values with respect to our place in the universe, thus calling for a re-evaluation of the assumptions behind varying positions to nature and to each other.

In *Green Speculations*, Eric Otto explores the shape of a form of radical ecology he calls transformative environmentalism, which combines influences from a diverse range of oppositional politics that have emerged since the 1960s, the science of ecology, environmental philosophy, deep ecology, ecofeminism and ecosocialism. He argues for 'environmental science fiction's place as a body of literature that reflects, sometimes prefigures, and in its finest moments theorizes transformative environmentalism and its assorted targets of criticism' (Otto, 2012, 4–5). Otto considers the different responses to environmental degradation that sf has offered, highlighting the prominent belief in the 'Illusion of Disembeddedness' from nature and its converse, 'part-of-nature' thinking, which builds on the implications of ecological interconnectedness. He also investigates different modes of environmental activism as a response to this degradation, alternatives that are influenced by sometimes contrary feminist approaches and by deep ecological thinking. Arguing that 'estrangement, extrapolation, and sense of wonder constitute an ecorhetorical strategy for works of fiction and nonfiction whose interests lie in questioning deep-seated cultural paradigms' (16–17), Otto identifies capitalism and its logic of limitless growth as the agent of this environmental degradation and the target of transformative environmentalism's critique.

In order to draw out the range of influences feeding into portrayals of terraforming, Lovelock's Gaia hypothesis needs to be considered in relation to the motif. The Gaia hypothesis claims that the Earth's planetary environment and its organisms are fundamentally interconnected in a biogeochemical cycle and that life provides feedback that assists in regulating Earth's climate (Lovelock, 1987). Although the implications of this paradigm for terraforming begin to cohere in the late 1980s, the Gaia hypothesis is central to the way terraforming develops during the 1970s. Ernest J. Yanarella argues that these two themes reflect each other and that terraforming is the Jungian shadow of the Gaia hypothesis (2001, 225–88). Lovelock suggests that terraforming Mars would be an 'unremitting task of nurture and the daily guidance of the newborn planetary life until it could, by itself, sustain homeostasis,' thus implying that terraforming would fulfil the reproductive criterion of a Gaian planet seen as a living organism (1995, 189). He goes on to write that '[t]houghts of Gaia will always be linked with space exploration and Mars, for in a sense Mars was the birthplace of the theory' (189).

In addition to the images associated with space exploration that have been informed by sf, Lovelock refers to his initial inspiration for the Gaia hypothesis, which he traces back to his work developing methods for detecting extraterrestrial life on Mars at NASA. Lovelock also collaborated with Michael Allaby to write *The Greening of Mars* (1984), which he explains inspired three scientific meetings, during one of which Robert Haynes 'coined the word ecopoiesis – literally, "the making of a home" – for the practice of transforming an otherwise uninhabitable environment into a place fit for life to evolve naturally' (Lovelock, 1995, 174–75). Ecopoiesis is a terraforming process that often appears in speculative scientific accounts and in sf as an early stage of a more comprehensive project. Haynes writes that 'the term refers to the fabrication of a sustainable ecosystem on a currently lifeless, sterile planet, thereby establishing a new arena in which biological evolution ultimately might proceed independent of further human husbandry' (1990, 180).

A Disciplined Thought Experiment: Landscaping, Sf, and Terraforming

> As science now approaches the 'how' of terraforming, science fiction must continue to explore the 'why'. (McKay, 1982, 309)

Sf narratives of terraforming offer imaginative spaces for reflection on fundamental issues regarding our place in relation to Earth, the planets of the solar system and the universe, reflection that in turn feeds into our practical attitudes and behaviour towards those spaces. Scientists and environmental philosophers have used the concept of terraforming as a thought experiment to consider human relationships to environments undergoing change. Daly and Frodeman cite Haynes's assertion that 'such a grand experiment would yield valuable information about the complex interworkings of ecosystem processes on Earth,' thus highlighting the experimental and scientific knowledge component involved in such thought experiments (2008, 145). This theme is prevalent in terraforming narratives and is often associated with the transformative image of a pastoral garden in contrast to desert spaces: in Pamela Sargent's *Venus of Dreams*, Iris Angharad explains that '[m]aking deserts here [on Earth] green again is not going to seem such a problematic undertaking if we can make Venus bloom' (1989a [1986], 237). Philosopher Robert Sparrow uses terraforming as a construct for the exploration of an agent-based virtue ethics, justifying his choice of scenario by arguing that '[t]he sheer scale of such a project allows many issues which arise around other modern technologically oriented environmental projects to be writ large' and that it demonstrates 'a shocking moral bankruptcy at the heart of our attitude toward the environment' (1999, 227, 229–30). Sf, as a literature of ideas that uses thought experiments to raise the idea or the imagined world to the status of hero, affords an ideal mode for speculative enquiry. The capacity for terraforming to be used to magnify issues connected to technologically based environmental projects and to examine the moral shortcomings that give rise to ecopolitical conflict makes it suited to contemporary environmental philosophical speculation.

Fogg explains that '[i]f one imagines the playground for thought experimentation as being a multidimensional space controlled by as many parameters as there are dimensions, it can be appreciated that, without any limits on the values of the parameters, the space can enfold an infinite number of possibilities' (1995, 88). To distinguish scientific enquiry from sf, Fogg describes a reduction 'to a subset of *real* possibilities' and the 'accept[ance] first and foremost [of] the constraint of physical law' (88). This book does not investigate the boundaries between science and sf, and nor does it read sf against '*real* possibilities.' Fogg's description of the conceptual space of the scientific thought experiment suggests an avenue for the literary analysis of representations of space.

Michael J. McDowell has argued that Mikhail Bakhtin's theories of dialogism and the chronotope incorporate 'much of the thinking about systems and relationships long ago embraced by the hard sciences,' thus offering ways to analyse texts from ecological perspectives (1996, 372). Murphy connects dialogism to Bakhtin's notions of 'answerability,' which 'represents the necessity of our responsiveness to be ethically grounded and morally justifiable,' and transgredience, the imaginative assumption of an external position to ground the writing and evaluation of artistic representations of nature (2011, 156). Murphy explains that transgredience 'encourages authors and critics to see themselves through another's perspective: those of the rest of the natural world at the general level, and of specific ecosystems, plants, or animals at the particular level' (156).

The chronotope as metaphorically applied to literature is a borrowing from the mathematical discourse underlying Einstein's theory of relativity. Bakhtin describes the *'chronotope* (literally, "time space")' as 'the intrinsic connectedness of temporal and spatial relationships that are artistically expressed in literature' (2002, 84). The way time and space are 'expressed' and how time is used to qualify spatial meanings and vice versa is central: language is used to give voice to the landscape by representing it as a site of traditional symbolic value as well as a space for the interaction of differing discourses. These values are represented synchronically, as textual spaces that are juxtaposed. Examples include the contrast between the icon of the domed city in Sargent's *Venus* trilogy and the Venusian environment visible outside of these domes. This environment is revealed diachronically through its representation within the text. Narration and character dialogue allow these spaces to come into contact with one another to construct a network of positions towards value systems within the text.

Darko Suvin draws on Bakhtin's concept of the chronotope for his adapted definition of sf, which he defines as 'a literary (etc.) genre *defined by the interaction of estrangement and historical cognition, and whose main formal device is a narrative chronotope and/or agents alternative to the author's empirical world*' (2008, 116, emphasis in the original). The chronotope as a unit of analysis collapses two central parameters of the sf thought experiment (space and time) with a third, the iconicity of particular spaces in textual representations of terraforming. Terraforming is an especially relevant motif with regard to the concept of the chronotope because it collapses textual world-building (imagined worlds) with representations of physical world-building. In Bakhtinian terms this involves the construction of a global chronotope (the planet to be terraformed) within a dialogic text that puts into play multiple interacting voices and their relationships to the environment. As the global chronotope can itself be broken down into a series of nested spaces at ever-decreasing scales, the fields for articulating various positions span a variety of continental, regional and local spaces.

The dialogic element of the chronotope usefully coheres with Damien Broderick's discussion in *Reading by Starlight* (1995) of the sf text's propensity to engage in a megatextual dialogue with other texts and discourses. Sf has developed a series of codes and reading protocols through the linguistic and discursive construction of sf icons, tropes and narrative trajectories. The sf megatext encompasses multiple sf discourses that have been constructed between texts and reader engagement, and its elements are open to considerable reformulation. By reconfiguring these icons a statement is made that resonates with the associations built up by the sf megatext. New meaning, icons and dialogues can thus be created by texts that engage with the sf megatext.

The environmental philosophical concept of landscaping emphasises that intentional modification of physical space involves an anthropocentric projection of cultural values onto nature. Bakhtin's discussion of the chronotope can thus be seen as a form of landscaping. Simon Hailwood defines 'landscape' as 'nature insofar as it is modified and interpreted for *human oriented ends*, moulded and used, or viewed as malleable and useful, for human interests and needs' (2007, 132–33, emphasis in the original). Landscaping thus involves both physical and intellectual processes: 'landscap*ing* [is] the ongoing historical process through which humanity physically shapes its environment[,] fills it with symbolic meaning, historical and aesthetic significance, and so makes itself at home' (Hailwood, 2007, 133, emphasis in the original). Hailwood extends Holmes Rolston III's original concept to include 'any discernible item assigned a symbolic significance within a culture' and notes that '[a]lthough physical and intellectual aspects of landscaping are distinguishable, they are not entirely distinct. The process is dialectical: possibilities of further action, modifications and interpretations are conditioned by those already in place' (2007, 133). Drawing on Karl Marx's note in *The German Ideology* regarding the socially constructed nature of landscape and on Gary Lease's notion in *Reinventing Nature?* that '*humans and nature exist in a dialectical relationship, each imagining the other,*' Hailwood explains that '[i]n landscaping we interact with nature and each other to create and transform the "material conditions" of human life and culture and so recreate ourselves' (133, emphasis added by Hailwood). In relation to this concept of landscape, Hailwood uses 'the term nature to mean nonhuman nature – *nature insofar as it is not landscaped* – although, of course humanity and human landscapes remain part of [...] nature in the all-encompassing sense of "everything"' (133, emphasis in the original). The concept of nature that Hailwood describes is cosmological (nature as everything) but exclusive of the human.

Representations of literary spaces in terraforming narratives, such representations being analysable as chronotopes particular to the discourse of sf, are examples of intellectual landscapes that construct imaginative spaces where social, political and ethical reflection on their implicit values can be explored. This focus on landscape is important because, as Dan McArthur argues, '[w]e will take our human moral environment with us to other worlds along with our pith helmets' (2001/2002, 13). Given that our contemporary attitudes and perspectives will guide our actions, McArthur argues that examining our ethical position with a view to evaluating these actions is of paramount importance. Like the literary notion of the chronotope, the concept of a moral landscape can be considered a specific form of Hailwood's notion of intellectual landscaping. Terraforming narratives are experimental spaces where political, ethical and aesthetic topographies, developed in a dialectical relationship between culture and Earth's nature, are overlaid.

The Lay of the Land

Who speaks for the land and for our relation to it? Who has the right to evaluate, judge and initiate a terraforming project that would alter a whole planet and have repercussions for its inhabitants and those involved in and affected by the economic, social and political relationships between that planet and Earth? Terraforming narratives often entail a consideration of economic, social, political and cultural relationships and strategies for negotiation and decision making. Issues of voice and the legacy of colonial history are central to their subject matter, which, through different permutations within and between texts, establishes spaces that allow political-cultural issues to be expressed and examined. Ursula K. Heise notes that considerations of place have become increasingly concerned with issues of globalisation, postnationalism and cosmopolitanism and argues for what she defines as an eco-cosmopolitan perspective towards a range of human politics of place that can embrace the 'more-than-human-world' (2008, 60–61).

Terraforming stories are underpinned by a will to transform planets according to a predetermined vision, often one that is homeworld-centric, in that new planets are terraformed against a blueprint derived, most frequently, from ecosystems on Earth. This follows from the definition of terraforming, since its proposed goal is to adapt planets to colonising peoples. *The War of the Worlds* (2004 [1898]) exemplifies the inversion: the red weed transforms Earth into a new Mars in an act of areoforming. Alternatively, and more in line with contemporary scientific knowledge regarding the possibilities for terraforming, planets are transformed according to the opportunities and constraints inherent in their environments: a compromise between the alien and indigenous is reached, often reflected by the entrance of pantropy into the narrative, as in Robinson's *Mars* trilogy.

Chapter one, 'Landscaping Nature's Otherness in Pre-1960s Terraforming and Proto-Gaian Stories,' examines how terraforming and proto-Gaian living world stories engage with environmental philosophical concepts. These concepts are based on ideas of nature's otherness and establish the philosophical discourse for the literary critical exploration of humanity's relationship to space adopted in this book. Nature's otherness refers to the relationship between humanity and features of the external world, and as such it intersects theories of mimesis and constructivism. Hailwood argues that nature's otherness need not be 'alien' or 'unfamiliar,' and he defines the value of nature's otherness as extrinsic, relational, non-instrumental and objective (2004, 35). He argues that respecting nature's otherness involves recognising its autonomy and its teleonomy. These two concepts are taken from philosopher Keekok Lee's three axioms: the Asymmetry, Autonomy and No-Teleology Theses. The assumptions underlying these theses can be unpacked with reference to Val Plumwood's examination of the hierarchical logic of dualism, which she argues corresponds to classical propositional logic. This hierarchical structuring has served 'to "naturalise domination," to make it part of the very natures and identities of both the dominant and subordinated items and thus to appear to be inevitable, "natural"' (1993, 32).

One of the reasons Plumwood analyses the historical bases of dualism is that 'the ancient forms do not necessarily fade away because their original context has changed; they are often preserved in our conceptual framework as residues, layers of sediment deposited by past oppressions' (1993, 43). She presents five operations by which 'dualism, the construction of a devalued and sharply demarcated sphere of otherness,' is conducted: backgrounding (denial), radical exclusion (hyperseparation), incorporation (relational definition), instrumentalism (objectification) and homogenisation or stereotyping (41). Instrumentalism in terms of value theory corresponds to Plumwood's instrumentalising dualistic operation; both contravene the Autonomy Thesis. Extrinsic worth is compatible with a relational definition, although in Hailwood's sense it does not involve the backgrounding of nature's otherness or its radical exclusion (nature's otherness can feature as elements of human landscapes).

Nature's otherness is represented in texts via depiction of non-human others, either as abiotic or as biotic forms of nature. Exploring the place of terraforming in the scientific romances of H.G. Wells (*The Shape of Things to Come*, 1967 [1933]) and Olaf Stapledon (*Last and First Men*, 1966 [1930]; *Star Maker*, 2004 [1937]), and in John Russell Fearn's pulp sf story 'Earth's Mausoleum' (1935), the first section of the chapter views nature's otherness in cosmological terms and connects it to the sublime. The section 'Pre-1940s Proto-Gaian Living Worlds' shows how nature's otherness is developed in two further works of scientific romance, M.P. Shiel's *The Purple Cloud* (2004 [1901]) and Arthur Conan Doyle's 'When the World Screamed' (2000 [1928]). Comparing these two stories to Edmond Hamilton's 'The Earth-Brain' (1936 [1932]), Jack Williamson's 'Born of the Sun' (1934) and Laurence Manning's 'The Living Galaxy' (1934) helps illustrate the shift in emphasis from the Burkean and Kantian sublime to sf's technological sense of wonder in 1940s pulp sf. The last section of this chapter considers the decline of the proto-Gaian theme during the postwar period by examining Murray Leinster's 'The Lonely Planet' (1949). These depictions of alien otherness draw on ideas of the sublime and grotesque to explore alternative conceptions of nature when viewed in a cosmological sense.

Chapter two, 'The American Pastoral and the Conquest of Space,' is divided into three sections that explore how sf adapts tropes and structures analogous to the Elizabethan pastoral's compression of meaning. The first, 'The Garden of the World in Early 1950s Terraforming Stories,' examines how Ray Bradbury's *The Martian Chronicles* (1958 [1950]), Robert Heinlein's *Farmer in the Sky* (1967 [1950]) and Arthur C. Clarke's *The Sands of Mars* (1976 [1951]) draw on the intersection between colonialism and the American pastoral to explore ideas of terraforming and independence from Earth. 'The Burden of Hope in the Garden of the Chattel: 1950s Consensus Dystopias,' shows how Frederik Pohl and C.M. Kornbluth's *The Space Merchants* (1974 [1953]), Walter M. Miller's 'Crucifixus Etiam' (1973 [1953]) and Poul Anderson's 'The Big Rain' (2001 [1954]) and *The Snows of Ganymede* (1958 [1955]) invert the optimism of space colonisation evident in Heinlein's and Clarke's texts by drawing on pastoral and dystopian imagery to depict the failures of colonisation.

Encounters with extraterrestrial life, while not an essential component, often intersect with the terraforming motif. Inhabited worlds pose significant philosophical problems for terraforming alien planets. These concerns are not absent from Bradbury's, Heinlein's and Clarke's texts, but they begin to receive a distinctly different emphasis by the late 1950s. Daly and Frodeman note that discussions of natural spaces often turn on questions of intrinsic and instrumental value, and that a focus on intrinsic value has tended to lead to 'ethical extensionism,' which 'depends on human definitions of moral considerability, which typically stem from some degree of identification with things outside us' (2008, 140). Identification with nature conflicts with a respect for nature's otherness that is based on recognising difference, but it does not completely exclude viewing nature as other to humankind. Section three, 'Moral Extensionism in Terraforming Stories of the Late 1950s and Early 1960s,' examines the shift in the ethical insights developed by the early terraforming narratives' investment in the exploration of cultural and political environments, which is extended to the landscape and the alien otherness of the indigenous populations. This section links aspects of the pastoral to environmental philosophical speculation in Anderson's 'Sister Planet' (1960 [1959]) and Clarke's 'Before Eden' (2001 [1961]).

'Ecology and Environmental Awareness in 1960s–1970s Terraforming Stories,' chapter three, is split into two parts that build on the themes of chapters one and two. The influence of the counterculture on sf transforms the environmental engagement of terraforming narratives during this period, introducing a wide range of ideas from mystical-spiritual engagements with nature, increased interest in communards, and the development of pragmatic environmentalism and appropriate technologies. The first section, '1960s–1970s Proto-Gaian Living Worlds,' considers the mythic-poetic register of proto-Gaian and terragouging narratives of the 1960s–1970s. Stories such as Richard McKenna's 'The Night of Hoggy Darn' (1964 [1958]) and 'Hunter, Come Home' (2001 [1963]), James White's 'Major Operation' (1971a) and 'Meatball' (1971b [1969]), Stanislaw Lem's *Solaris* (2003 [1961]) and Ursula K. Le Guin's 'Vaster Than Empires and More Slow' (1982 [1971]) and *The Word for World Is Forest* (1976 [1972]) explore colonial themes in the context of the living world as a counter to exploitative relationships to the land. Lem's *Solaris* draws from the tradition of philosophical speculation pioneered by Stapledon to offer a fundamental critique of the colonial and scientific project of discovery and thus departs from the approaches of the other authors considered in this section. Although published in Polish in 1961, it was first translated into English in 1970, and so *Solaris* is discussed in the context of its English translation. Le Guin's *The Word for World Is Forest*, while not a living world story, deploys the chronotope of the forest in ways that chime with many of the other stories in this part, and so examining it as a work of terragouging offers fruitful insights into the development of the terraforming narrative.

The discussion then shifts in the second part to an analysis of ecology and politics in Frank Herbert's seminal *Dune* trilogy (1965; 1971 [1969]; 1976), before moving on to examine Robert Heinlein's *The Moon Is a Harsh Mistress* (2001 [1966]) and Le Guin's *The Dispossessed* (1999 [1974]) and their exploration of libertarianism, anarchism and independence as alternatives to the instrumentalising imperialism of *Dune*. This chapter ends by considering the significance of Ernest Callenbach's *Ecotopia* (1978 [1975]) to ecological and environmental sf. The texts in this part both influence and draw from the mixture of technologism and environmentalism that coalesced around Stewart Brand's *Whole Earth Catalog* (1968), a publication that was connected in multiple ways to sf. The link between politics, the pastoral and terraforming first shaped by the works of the 1950s terraforming boom undergoes a transformation in the light of countercultural influences and the popularisation of ecologism and environmentalism.

Chapter four, 'Edging Towards an Eco-cosmopolitan Vision,' examines transformations of the terraforming motif that are rooted in the wider cultural impact of the modern environmental movement and the increasingly widespread popularisation of James Lovelock's Gaia hypothesis. Terraforming as an act of world-building resonates with the utopian impulse to remake socio-political worlds. It lies in tension with the urge to explore space and further extend human frontiers. Those who choose terraforming must stay and work to create a habitable planet and equitable society, participate in the continuation of damaging forms of socio-political structures or preserve the planet in as much of its original form as possible. Those who leave must abandon these societies indefinitely.

This chapter considers Michael Allaby and James Lovelock's *The Greening of Mars* (1984), Pamela Sargent's *Venus* trilogy (1989a [1986]; 1989b [1988]; 2001a) and Frederick Turner's epic poem of terraforming, *Genesis* (1988), all of which feature characters who confront this dilemma. Although Sargent's first two instalments of the *Venus* trilogy were published in the 1980s, the final part was published in 2001, but it is nevertheless considered in this chapter. Like Lovelock, who greatly influenced wider scientific and environmental perspectives, Turner has had a direct influence on scientists and was one of the first writers to explicitly engage with issues of environmental philosophy, bioethics, the relationship between science and the humanities and space colonisation: his work was placed on recommended reading lists amongst NASA employees, he was involved in the founding of the Mars Society and he was invited to the 1991 terraforming workshop, where he collaborated with scientists such as Robert Zubrin, Carl Sagan and Martyn Fogg. These works articulate a sense of complexity and conflict that accompanies an awareness of the plurality of visions for the future. Terraforming narratives have become highly ramified since the 1980s. Awareness of the past, of traditions that offer resources for habitation and the making of homes on other worlds, are central to the representation of terraforming in sf.

The final chapter of this book, 'Kim Stanley Robinson's *Mars* Trilogy,' brings the themes explored in previous chapters to a consideration of the groundbreaking terraforming trilogy of the 1990s. The *Mars* trilogy continues the conscious reflection on past traditions that the dialogic aspect of sf has built upon throughout the terraforming tradition. This section connects ideas of landscaping to Jed Rasula's notion of 'composting,' an appropriate figure in the context of terraforming for the ramified dialogism of terraforming texts, which draws not only from sf but from scientific and environmental discourse (2002). Thierry Bardini, too, explores terraforming in the context of 'junk,' a notion compatible with composting that turns on bootstrapping, a basic concept for cybernetic conceptualisations of terraforming and ecopoiesis (2011). This chapter considers Robinson's engagement with pastoral themes, environmental philosophical perspectives, the influence of ecologically informed economics and the socio-political consequences of terraforming. This book ultimately shows how the terraforming motif is used to connect the values that underpin our modes of understanding and relating to culture and politics, habitation and the environment.

The assumption behind attempts to transform planets is that change is enacted for the improvement of society. The utopian tradition, like some sf, draws on pastoral and romantic treatments of the landscape that also feed into visions of terraforming. Terraforming allows for the convergence of multiple and often competing themes and positions in a single narrative that engages with political and ecological issues, thus making it central to sf. Fredric Jameson draws on Darko Suvin's connection of sf to utopia and Brechtian estrangement to claim that 'terraforming ought to constitute the utopian moment par excellence' (2000, 220). This link is given greater resonance in the context of geoengineering. When we allow that attempts to alter Earth's landscape are also aspects of terraforming, it becomes clear that terraforming touches on a major element of all human civilisations. The concept of landscaping relates to terraforming in a direct way under this broader definition. Robinson considers terraforming in this manner when, in the final autobiographical short story in *The Martians*, the narrator admires the trees planted alongside a Californian street and describes the scene as '[t]erraforming at its finest' (2000 [1999], 456). The ability to alter the landscape and the ethical dilemmas this poses direct attention towards the future. The fundamental question asked is how we want to live, and it emerges from the concern over whether we can continue living in ways that threaten the integrity of our environments.

1: Landscaping Nature's Otherness in Pre-1960s Terraforming and Proto-Gaian Stories

The 1960s saw the consolidation of a growing awareness of environmentalism. Rachel Carson's 1962 *Silent Spring* popularised the American environmental movement, while James Lovelock first conceived the basis for the Gaia hypothesis in 1965. Their writings influenced the developing environmental movement, if not always in ways that Lovelock approved of, yet prior to the 1960s sf had engaged with ideas that anticipated and fed into modern environmental philosophy and what Lovelock now refers to as Gaia theory. Anna Bramwell traces the growth of ecological awareness from its origins in the 1880s and notes how the scientific romances of writers such as H.G. Wells incorporated ecologically inflected concepts into their narrative explorations of the relationship between nature and politics. Bramwell differentiates between ecologism, which is concerned with global social and economic change, and environmentalism, which diverged from ecologism in the 1930s to focus on specific problem-solving at localised political levels (1990, 104–05).

Terraforming stories explore the otherness of nature, a view that opposes the interconnectedness without boundaries that deep ecology endorses. Links have been forged between holist approaches to the physical sciences raised by the Gaia hypothesis and the deep ecology movement. Val Plumwood explains that 'dominant forms of deep ecology choose for their core concept of analysis the notion of identification, understood as an individual psychic act rather than a political practice' (1993, 17). In its weaker formulation, recognition of nature's otherness can be considered a form of deep ecology when it is understood simply as an attention to 'the interconnectedness of the natural world (as in some sense the bearer of non-instrumental value) and humanity's place within it, so as to effect fundamental social change that will bring humanity to its proper place' (Hailwood, 2004, 36). Deep ecology's strong identification with nature, however, raises problems for recognising nature's otherness. Citing Bill Devall and George

Sessions's claim that '"to the extent that we perceive boundaries, we fall short of deep ecological consciousness"' (121), Hailwood argues that stronger forms of holism often associated with the Gaia hypothesis are incompatible with the recognition of difference involved in conceptions of nature's otherness (36).

This chapter examines terraforming themes during the interwar period in the work of two influential writers of scientific romance, H.G. Wells and Olaf Stapledon, and in British pulp writer John Russell Fearn's short story 'Earth's Mausoleum.' The second part examines the use of the living world motif in scientific romances by M.P. Shiel and Sir Arthur Conan Doyle, and in a cluster of American pulp stories by Edmond Hamilton, Jack Williamson and Laurence Manning, three significant shapers of the 1930s space opera. This discussion concludes by moving beyond the interwar period to consider how the themes connected to the living world motif were taken up during the postwar period by Murray Leinster. These works use the terraforming and living world motifs to explore orientations to nature's otherness, anticipating ideas later incorporated into the Gaia hypothesis.

Terraforming as a Site for Environmental Philosophical Reflection

Many thinkers have used terraforming as an experimental site for the exploration of science and philosophy. This approach trades on the notion that sf is a literature of ideas, offering textual spaces for thought experiments in the tradition of the *conte philosophique*. Erin Moore Daly and Robert Frodeman argue that space exploration and environmental ethics were generated from a closely related set of socio-economic and historical factors and that 'before the launching of Sputnik in 1957, philosophic consideration of space was lodged within the science fiction literature of H.G. Wells, Jules Verne, Ray Bradbury, Arthur C. Clarke, and others' (2008, 137). Daly and Frodeman outline the various ethical positions on the terraformation of both lifeless planets and planets that support microbial life. They show how this debate tends to orbit around notions of intrinsic value, the worth an object has by virtue of its inherent properties.

Both David Grinspoon and Christopher P. McKay support the introduction of life to otherwise uninhabited planets, while Robert Zubrin takes a stronger position by supporting the introduction of life to planets already inhabited by microbial life. Holmes Rolston III contests this view and argues that the habitation of lifeless planets should be avoided, while another preservationist, Alan Marshall, strictly opposes any sort of interference whatsoever. Martyn Fogg counters this view by pointing out that such arguments 'are often misanthropic critiques of human nature' or 'sentimental illusions based on out-of-date ecology'; he argues that abiotic nature cannot have intrinsic value because it cannot think, act, or care (Daly and Frodeman, 2008, 146–47, 147). While Robert Sparrow notes the 'shocking moral bankruptcy' that terraforming scenarios highlight, his arguments do not preclude the possibility of terraforming, but rather 'suggest that we examine ourselves and how the project reflects on our character before we undertake it' (1999, 239). He explains that '[i]f for instance, terraforming were a project undertaken with genuine reluctance, in full knowledge of what was being destroyed, because no alternative existed for the survival of the human race,' such projects would not demonstrate hubris or aesthetic insensitivity and would therefore be a legitimate, if regrettable, action (240).

Daly and Frodeman point out that while abiotic nature may not have intrinsic value, 'the *relatedness* of nature and natural objects to human beings' may be grounds for a relational, extrinsic value: '[w]e may be confident that rocks do not think, or have values of their own. But humans can nonetheless value rocks for their own sake – they can be experienced as beautiful, sublime, or sacred' (2008, 147, emphasis in the original). These aesthetic categories include important responses to abiotic nature and have traditionally provided an arena for environmental philosophical debate for thinkers such as Allen Carlson (1979; 1981; 2002), Stan Godlovitch, Yuriko Saito (1998), Thomas Heyd and Patricia M. Matthews (2001). Nevertheless, there are grounds for arguing that Simon Hailwood's concept of nature's otherness provides for such an extrinsic, relational value to be accorded to both biotic and abiotic nature without relying on appeals to aesthetic value. Hailwood argues that nature, albeit a social construction, also possesses an aspect that cannot be reduced to the social sphere, what he calls 'nature's otherness.' Although Hailwood notes that nature is present '"from the streetcorner to the stratosphere"' and that nature's otherness is therefore a property of familiar, unfamiliar and alien examples of nature (2004, 35), Godlovitch summarises the essence of the strong form of the relationship between humankind and nature when he contends that 'Nature is, for us, fundamentally inaccessible and alien' (1994, 19). Sf's investment in radical, alien otherness makes this notion useful for critical consideration of specific works from environmental philosophical perspectives. Through the deployment of iconic sf others, such as the robot, the alien and the adapted human, sf has engaged with the politics and philosophy of otherness from multiple theoretical positions. In contrast to human-centred critiques such as Marxism, feminism, and postcolonialism, ecocritical perspectives interrogate anthropomorphism and can be seen as anti-humanist in McKay's sense of the term: '[h] umans have no special rights or needs and do not determine the value of the environment' (1990, 188). Nevertheless, there are significant intersections between these positions, and in much sf they often work in mutually supportive ways.

Keekok Lee uses the example of terraforming to examine nature's value when nature is understood to encompass other parts of the solar system. She notes that many environmental ethicists find it 'deeply problematic to argue that abiotic Nature could have intrinsic value', because this is a property more often accorded to biotic life (1994, 90). Addressing McKay's observation that environmental axioms tend towards a narrowly Earthbound environmental ethics, Lee uses the example of terraforming because 'an environmental ethics informed by features unique to Earth may be misleading and prove inadequate as technology increasingly threatens to invade and colonize other planets in the solar system' (89). She begins her thought experiment by proposing that 'Mars has no organic life and, therefore, it follows, no intrinsic value' (91). Developing this line of thought, Lee considers that it may be morally permissible to produce biotic life in places where it has undergone degradation; from here it is a short but unwarranted leap to the rhetorical question, 'if such permission [for nature's restoration] obtains on Earth, why not Mars?' (91). Lee explores the necessity for an environmental ethics that 'confront[s] the issue of abiotic or inanimate Nature as a locus of intrinsic value' (92).

The War on Nature in Wells's *The Shape of Things to Come* and John Russell Fearn's 'Earth's Mausoleum'

The interwar period hosted a debate between two strands of ecologism. The first is a monism that can be traced back to Ernst Haeckel, who coined ecology in 1866, and his student Hans Dreisch, who popularised ecological vitalism in Germany and in a series of lectures in Scottish and English universities until 1913. The other strand derives from energy economics, which is based on the recognition of entropic processes and the finiteness of Earth's resources, ideas often understood as implying a mechanistic view of nature. Bramwell argues that Wells and Stapledon drew from the energy economics prevalent during the interwar period and cites Wells as an example of the scientific utopians who preferred strategies of global planning conducted by a scientific elite (1990, 54, 65). Wells was a member of the Fabian Society, an organisation that shared with the Monist League (which inspired Haeckel's ecological vitalism) Wells's trust in the utopianism of a cadre of scientists who could bypass the contemporary political process.

In *Vibrant Matter*, Jane Bennett considers the philosophical history of vitalism and sketches a framework for a contemporary vibrant materiality that possesses a distributive agency. By 'vitality,' Bennett refers to 'the capacity of things – edibles, commodities, storms, metals – not only to impede or block the will and designs of humans but also to act as quasi agents or forces with trajectories, propensities, or tendencies of their own' (2010, viii). It is important to consider the vibrant materiality of things because 'the image of dead or thoroughly instrumentalized matter feeds human hubris and our earth-destroying fantasies of conquest and consumption' (ix). While Bennett focuses her analysis on non-living matter, on things that, she argues, possesses an 'active, earthy, not-quite-human capaciousness' (3), she invokes Vladimir Ivanovich Vernadsky's early formulation of the notion of the biosphere and his emphasis on the continuity of biotic and abiotic matter (8). She writes that 'in the long and slow time of evolution, then, mineral material appears as the mover and shaker, the active power, and the human beings, with their much-lauded capacity for self-directed action, appear as its product' (11).

Wells's *The Shape of Things to Come* (1967 [1933]) is a future history that recounts an episode of 'geogonic planning' during the world state's technocratic management of Earth. This example of terraforming is a form of geoengineering embarked upon after a long period of war and global unification leads to a scientific utopia. Wells sent a copy of this work to his correspondent and admirer Olaf Stapledon, whose *Last and First Men* (1966 [1930]) and *Star Maker* (2004 [1937]) utilise the future history within a cosmic schema as part of their design as 'essay[s] in myth creation' (Stapledon, 1966, 12). Brian Stableford explains that Stapledon's works 'construct imaginary worlds to embody metaphysical theses' (1985, 138). Stapledon incorporated into *Last and First Men* elements of J.B.S. Haldane's 'The Last Judgement' (1927), which proposes the terraformation of Venus, combining this with J.W. Dunne's speculations on alternative temporalities and prescient dreaming in *An Experiment With Time* (1981). Unlike Wells's work, which showed a reliance on energy economics, Stapledon's works are indebted to vitalist philosophies that emphasised notions of a life-force underpinning conceptions of nature.

The Shape of Things to Come features episodes of geoengineering representing the culmination of civilisation's effort to address a deep-rooted anxiety towards nature. This anxiety underlies a complex of environmental relations and effects that appear in earlier phases of civilisation's development, such as the colonial appropriation of resources, international war and the dramatic reduction of species diversity and species' environments. At a late stage of the world state's growth, Earth is described in a way that justifies the physical mastery of the planet and confirms the socio-political system supporting the state: '[f]rom the air on a map it was manifest that the world was still "governed." The road system was like a net cast over a dangerous beast' (Wells, 1967, 442). The endeavour to govern nature by ensnaring it in a planetary road system is symptomatic of humanity's fundamental dependence on a hostile environment.

Such dependence is portrayed in the chapter '1933: Progress Comes to a Halt,' where the narrator highlights a representative instance of 'a terrific defeat for [a town near Cardiff] in the war upon Nature' (Wells, 1967, 132). This event, in which a local mine explodes and kills three hundred people, stands as a microcosm for humanity's dependence on and subjugation to natural forces. The text's fictionalisation of the difficulty of recovering economic and social wellbeing in the wake of the Great War of 1914–1918 enlarges the scope of humanity's vulnerability to nature. These local and international events exemplify civilisation's physical fragility. The threat symbolised by the emblem of Earth as a dangerous beast signifies an uneven human relationship to nature. As a consequence of humanity's 'war against nature,' civilisation's changing technological abilities allow it to mitigate this dependence on and vulnerability to its environment.

The emblem of the ensnaring road system is a landscape that symbolises progress. The world state's political unification and technological proficiency is seen as the solution to humanity's difficult adaptation to nature. Roads overcome spatial constraints and unify disparate nations by facilitating travel and communication over long distances. Contrasting the old order to the saner technological utopia of the 'Modern World State', the narrator identifies a disparity between two forces as the underlying reason for the former's failure to overcome both nature and civilisation's fragility: '[t]he great processes of mechanical invention, which have been described in our general account of the release of experimental science from deductive intellectualism, were increasing the power and range of every operating material force quite irrespective of its fitness or unfitness for the new occasions of mankind' (Wells, 1967, 48–49). Deductive intellectualism is essential to direct the use of new technologies in accordance with the needs of the time, lest unchecked use of power steer civilisation to war and socio-political collapse. During the ascendancy of the Modern World State the dialectic between humanity and its environment is subject to civilisation's increasing influence, allowing humanity to overcome the failures of the old socio-political order. The application of technologies capable of altering the planet's environments offers opportunities for directed human evolution; the narrator speculates that 'an increase in desirable habitats may bring with it an increase in the variety of desirable human types' (Wells, 1967, 455). The dialectic underlying civilisation turns on evolution, which governs the relationship between the human and the non-human and which is in turn connected to ideas of progress, 'the essential and permanent conflict in life between the past and the future, between the accomplished past and the forward effort' (54). The developing world state embodies this conflict as it attempts to redress the imbalances between technological capability and its fit use wrought by the excesses of the old order.

The Transport Control fails to effectively maintain global unity, eventually ceding to an enlightened scientific community capable of better managing Earth's resources. This saner, more reflective humanity is able to resolve many environmental and geopolitical problems. Recapitulating economist Henry George's metaphor in *Progress and Poverty* (1920 [1879]) of the Earth as a well-provisioned ship sailing through space, the narrator explains that this new government establishes a way of living that shows how '[t]his planet, which seemed so stern a mother to mankind, [was] discovered to be inexhaustible in its bounty. And the greatest discovery man has made has been the discovery of himself' (Wells, 1967, 487). This vision of nature remains instrumental, as it is considered both a resource and an object for autological speculation, which reflects concern onto the study of the human self.

The socio-political sphere is mapped against nature through its structuring as a metaphorical 'ecology' (Wells, 1967, 290). As humanity is better able to manipulate the planet, '[h]istory becomes a record of increasingly vast engineering undertakings and cultivations, of the pursuit of minerals and of the first deep borings into the planet. New mechanisms appeared, multiplied, and were swept away by better mechanisms. The face of the Earth changed' (Wells, 1967, 413). Despite these changes and the early exploration and mapping of 'the last *terra incognitae,*' Earth for the Second Council was 'a world of promise still to be fulfilled' (Wells, 1967, 51, 445). An early anticipation of geoengineering appears in the brief chapter 'Geogonic Planning,' where the '[m]odification of the planet-levels operating in conjunction with the restoration of forests now in progress' is considered (454). Here, the world state's plans for planetary engineering remain a dream of 'moulding a fire-sprouting, quivering planet closer to the expanding needs of man'

(455). Geogonic planning represents a stage in the development of civilisation in which 'deductive intellectualism' directs the use of technology, thus allowing humanity to alter nature to provide ideal foundations for a scientific utopia whose expansion is charged with colonial ideology. In either case the form of progress that is advanced here, evident from the landscaping analysed by the narrator, is a response to an asymmetric relationship between humanity and non-human nature that disrespects nature's autonomy. Nature in this context is Earthbound and non-human. Human relationships to nature tend towards instrumentalism, with nature functioning as a resource for and background to civilisation. Human dependence on this beastlike non-human nature exacerbates the dangers that nature poses to civilisation, imagined as a sharply demarcated entity. This understanding of nature attempts in part to account for the failure of 'The Age of Frustration' to realise its project of progress.

The Shape of Things to Come was adapted in 1936 into the well-received film *Things to Come*, which popularised the broad themes of Wells's future history. It portrayed the decline of civilisation after war and the struggle from 'barbaric' nation states to a scientific world civilisation riddled by widespread disaffection with the dominant ideology of progress. This film stages a debate over the worth of science for society, and it retains a sense of the asymmetry between humankind and nature. The rise of the scientific age is preceded by a period of rebuilding described as 'an active and aggressive peace' where all of Earth's resources are tapped to 'put the world in order' (Menzies, 1936). Theotocopulos, the leader of a rebellion against continuing progress in the year 2036, concedes that this age of 'machines and marvels' has conquered nature and built 'a great white world' of artificially lighted underground cities, thus replacing nature's threatening otherness with a construct rigidly controlled by technology (Menzies, 1936). The conflict in this future age centres on a rebellion against progress that is expressed as a desire for the destruction of a 'space gun,' the world civilisation's first attempt at interplanetary travel and a symbol of continuing scientific progress (Menzies, 1936). Theotocopulos's arguments centre precisely on the Promethean fear that cosmological nature engenders: anticipating further journeys into space that the success of the space gun portends, Theotocopulos foretells in a public speech that 'the time will come when you in your turn will be forced away to take their chance on strange planets, on dreary, abominable places beyond the stars' (Menzies, 1936).

This ideological opposition centres on the idea of the colonisation and shaping of other worlds in the future, a possibility that is ultimately endorsed by the film's climax. Contesting notions of progress with ideas of freedom, rest and safety, Theotocopulos argues that the sacrifice of human lives to technological and scientific progress will eventually alter the face of the whole world once again, thus forcing the majority who are content with current circumstances to adapt to new conditions. These primarily social concerns are symbolically invested in a confrontation with untamed nature. The conquerors of the old world order, who directed their mastery towards other societies, are replaced by a government that attempts to dominate first local and then increasing scales of nature, before this tendency culminates in an expansion outward towards space. Oswald Cabal, the grandson of the legendary John Cabal, who inaugurated the scientific age, argues that any life worth living must continue to advance, and that this advance can only be achieved by facing death, symbolised by space. Only by striving for a continual expansion outward to the stars and thus confronting human limits can living be made worthwhile.

Lee's Asymmetry Thesis claims that while humans are dependent on nature, nature is not similarly dependent on humanity: '[n]ature's own existence and functioning integrity is independent of human existence' (Lee, 1994, 93). The Autonomy Thesis builds on this by stating that 'the Earth and its extremely complex biosphere are fully autonomous' (93). Lee uses the example of human extinction to point out that nature would continue to operate in the absence of humanity. She defines autonomy here as nature's 'ability to exist, to function integratively and well without any reference to, assistance from or reliance on humans' (93). *The Shape of Things to Come* and *Things to Come* recognise the Asymmetry Thesis by demonstrating humanity's dependence on Earth while reflecting on Earth's independent existence, although they do illustrate nature's physical fragility through its devastation by war and humanity's adaptation of its surface. They also show a disrespect of the Autonomy Thesis and nature's otherness in that Earthbound nature is only valuable instrumentally: as a condition for ensuring human comfort and as a space where humanity can, narcissistically, reflect on its own mind. While this is not in itself problematic, these works take the extra step of disregarding the way these human landscapes fail to recognise nature's autonomy and otherness, thus contributing to an attitude that leads to its domination. The emblem of Earth as an imprisoned beast and the theme of civilisation's war against nature are reactions against this asymmetry and responses that curtail nature's autonomy. Hailwood suggests that nature's otherness can be usefully predicated on the Autonomy and No-Teleology Theses (the latter discussed below), but that the additional Asymmetry Thesis could lead to a Promethean fear that would justify the domination of nature in order to counter its threatening aspect (2004, 31). This dynamic is amply demonstrated by the notion of progress and its entailments depicted in *The Shape of Things to Come* and its cinematic counterpart.

John Russell Fearn's 'Earth's Mausoleum' (1935) is conspicuous as one of few stories published in the 1930s that centres on planetary adaptation, and many of its themes became staples of the terraforming narrative. In this story, the original impactor that separated the Moon from Earth re-emerges and is revealed to be an alien spacecraft. Its occupants are scientists who freely share their technology and expertise with humanity to organise and transform the Earth. They represent the ideal combination of deductive intellectualism in their capacity for scientific world management and experimental science in their superlative technological achievements. Crespin, a scientist and the appointed successor to the alien leader Mayro, sees terraforming as an extension of geoengineering, an act of conquest to 'extend Earth's ramifications' (Fearn, 1935, 72). Mayro directs a 'One-Year Plan' to open up Earth's environments and resources for human use, 'the vast improvement in the constructional scheme of the world' (71). Once their war on nature tames the Earth, the aliens turn their attention to the grand project of terraforming the Moon to 'cultivate a perfect little world' (75). Terraforming involves the construction of massive atmosphere generators and much human labour, leading one scientist to think that they are dealing in things too big for them, 'beating nature at her own game, so to speak' (75). As the iconic image of a 'softly green "new" Earth visible from the sky' of humankind's first planet takes shape, the second phase of terraforming begins as the tough community of pioneers level mountain ranges and crater walls to shape the Moon's topology (75).

Mayro's discovery of a method for liberating atomic force and the invention of a powerful 'dredger' triggers growing social unrest as the once happily busy colonists are deprived of fulfilment in hard work. Capitalising on this disaffection, saboteurs disrupt the dredger but, because they fail to appreciate how it operates, accidentally initiate a process that siphons energy from the Sun. To avert disaster, dreams of a utopian new Earth are abandoned when the Moon is converted into a new sun. Mayro and the surviving aliens, well-meaning scientists who underestimate the impact of introducing new technology and social arrangements to human society, sacrifice themselves and their ship so as to ensure that humankind survives the catastrophe. Like the scientific utopians of *Things to Come*, they fail to appreciate society's resistance to rapid and sweeping technological change. Terraforming, an endeavour that represents the best that science and technology have to offer, promises fulfilment and power to humankind only if human nature can accept the social changes it brings. 'Earth's Mausoleum' endorses the control that science and technology offers at the expense of respecting nature's autonomy, but it assigns the failure to respect nature's fragility to the saboteurs: their casual manipulation of technologies they do not understand shows a failure to acknowledge human dependency on both nature and technology.

Nature's Otherness and Terraforming in Stapledon's *Last and First Men* and *Star Maker*

Wells and Fearn depict human relationships to nature in terms of a universalising progress that remained widespread in sf, especially in terraforming and Gaian narratives. Stanislaw Lem would later attack this theme in *Solaris*, parodying it as 'the long historic march of humanity' (2003 [1961], 167). The war on nature theme appears in Stapledon's *Last and First Men* (1966), in which the eighth men's terraforming of Neptune is described as 'the story of man's attack upon his final home' (271). This attack, following the earlier terraforming of Venus, grows from a similar anxiety towards nature as that depicted in Wells's text, although nature in Stapledon's is non-human in a cosmological and not Earthbound sense. *Last and First Men* and *Star Maker* examine human nature from vast evolutionary perspectives, thus providing a vector for considering the various meanings attached to the concept of human nature. The metaphysical theses that Stapledon explores in these works open up spaces that converge with contemporary environmental philosophical speculation.

Stapledon's *Star Maker* engages in significant ways with speculation on cosmological nature's alien otherness. Metaphorical proto-Gaian images appear in *Star Maker*, resonating with Wells's references to Earth as a 'beast' and a bountiful mother. When Stapledon's narrator leaves Earth and sees 'a creature alive but tranced and obscurely yearning to wake,' he refers to an impression that would accrue symbolic resonance with images of Earth as Gaia (Stapledon, 2004, 15). Stapledon uses this image to speculate on 'awakened worlds,' where the intelligent population of a planet attains communal consciousness (271). Representations of stellar intelligences, stars that the narrator learns 'are best regarded as living organisms, but organisms which are physiologically and psychologically of a very peculiar kind,' also appear in the narrative (194).

Conversely, the figure of the Star Maker is quite different from these proto-Gaian images: its ability to create worlds and universes aligns it with terraforming; the Star Maker embodies the human dream of world-building seen as a godlike endeavour. Terraforming as godlike creation is central to many later terraforming narratives, and is here linked to

Stapledon's meditations on spacetime and origins. *Star Maker* magnifies the scale of spacetime portrayed in *Last and First Men*, and it extends the consideration of alien civilisations by using the device of a disembodied narrator who eventually merges with the consciousness of other aliens to form a pan-psychic cosmic mind. This cosmic mind's search for the Star Maker ends with a vision of a being of vast power which, over the course of many experiments in universe creation, matures to fulfil its potential. The narrator experiences the sublime when confronting the Star Maker; its presence in the text operates as an emblem of nature's otherness. The narrator's various theories about this being are attempts to encompass its alien otherness with intellectual landscapes.

Last and First Men chronicles two billion years of humanity's fluctuating development 'from savagery towards civilization' (Stapledon, 1966, 21, 276–77). Human nature for Stapledon is subject to the vagaries of the cosmos, to socio-economic, cultural and political influence and to philosophical ideas that underlie and provide the moral imperative driving civilisation's activity. Against changing physical environments Stapledon explores the beliefs that seem to justify – for each iteration of humankind – the oppression of both human and non-human others. He traces the intra-human and extraterrestrial conflicts and the ensuing crises that lead to key events determining humanity's future adaptation of nature. Stapledon also raises questions of dominance related to the mastery of nature that the development of progressively more advanced forms of technology offers to civilisation. As in Wells's book and film adaptation and Fearn's short story, the overarching philosophical vision explored is important – in the case of *Last and First Men* the later terraforming episodes function as extensions of earlier tendencies exhibited during episodes set on Earth. Stapledon traces the rise and fall of civilisations in a series of cycles to juxtapose each human species' character in abstracted form.

The narrator declares that *Last and First Men*'s aim is to 'help you to feel not only the vastness of time and space, but also the vast diversity of mind's possible modes' (Stapledon, 1966, 17). Such vastness operates as a conceptual contrast to humanity's relative insignificance, as a precondition for the variety of evolved human types, and as a signifier of the sublime. The threat that cosmological nature poses to humanity spurs its search for greater powers of manipulation over the physical universe. The narrator explains that '[Humanity's] existence has ever been precarious. At any stage of his career he might easily have been exterminated by some slight alteration of his chemical environment, by a more than usually malignant microbe, by a radical change of climate, or by the manifold effects of his own folly' (Stapledon, 1966, 314). Terraforming, seen in the context of a history of technologies designed for the purpose of controlling nature, is a technologically sophisticated answer to the Asymmetry Thesis. Wells's *The Shape of Things to Come* shares the theme of civilisation's fragility, prominent in later terraforming narratives and central to stories of ecotastrophe. This sense of human insignificance and fragility when compared to the vastness of cosmological nature leads to a sense of Promethean fear that raises issues pertinent to Lee's Asymmetry and Autonomy Theses.

Star Maker begins with reflections on cosmological nature and human insignificance that bear on the problem of an individual's sense of place within a cosmic schema: '[c]onsidered even without reference to our belittling cosmical background, we were after all insignificant, perhaps ridiculous. [...] Even the cold stars, even the whole cosmos with all its inane immensities could not convince me that this our prized atom of community, imperfect as it was, short-lived as it must be, was not significant' (Stapledon, 2004, 9). This cosmological indifference is set against a localised, human significance endued by the community of marriage, that 'intricate symbiosis' (Stapledon, 2004, 7). Recognition of nature's indifference thus stated is anti-humanist in McKay's sense, but not misanthropic; it is the first step towards a critical examination of the human individual and community and their place as part of an indifferent cosmological nature. In a reference to *Last and First Men* in *Star Maker*, this theme is again addressed in relation to the largest unit of community: '[a]ll this long human story, most passionate and tragic in the living, was but an unimportant, a seemingly barren and negligible effort, lasting only for a few moments in the life of the galaxy' (184). The total human community, a category inclusive of alien societies in Stapledon's work, is insignificant from the perspective of the cosmos, thus spurring civilisation's effort to overcome its limitations.

The sense of human insignificance motivates humanity's intervention with the physical world in *Last and First Men*: 'the Second Men were oppressed by the brevity of human life, and the pettiness of the individual's achievement in comparison with the infinity round about him which awaited apprehension and admiration' (Stapledon, 1966, 147). This ambivalent feeling of 'apprehension and admiration' for nature's otherness recognises a nature independent of humanity's needs and desires: non-human cosmological nature is fully autonomous and possesses an asymmetrical relationship to humanity. 'Apprehension' as perception and appropriation suggests that human responses to nature's otherness combine attempts to understand and control that nature with a respectful admiration predicated on a feeling of the sublime in response to infinity and vastness. Godlovitch suggests that a culture-inspired, centric natural aesthetic is necessarily 'accountable to and acceptable only within the bounds of human perception and human apprehension' and is therefore arbitrary (1994, 18). Stapledon's cosmic perspective allows multiple cultural approaches to nature to be explored while retaining a sense of the contingency and limits involved in landscaping nature.

In keeping with this focus on vast time periods, Stapledon depicts human societies as progressing through several permutations of pastoral, agricultural, and industrial stages, and through several types of governments, including medieval and cosmopolitan structures. The second men are envisioned at one point as 'noble savages, [who] then passed rapidly through the pastoral into the agricultural stage' (1966, 139). Each of these stages is marked by crises where development past a critical threshold depends on humanity overcoming its natural response to new technologies and socio-political clashes while flexibly adapting to the physical world and appropriately managing natural resources. Nations progress through periods of nationalist, regional, and then world-state governments and, towards the end of the text, a global society turns towards the vast project of colonising space and terraforming other planets. *Last and First Men* draws oppositions between industrial stages informed by instrumental values of progress versus utopian or idyllic stages, and between intellectual versus philosophical tendencies. Terraforming is often predicated on attempts to gain mastery over nature. In *Last and First Men*, Stapledon presents another motivation for planetary adaptation.

The fifth men terraform Venus in order to escape the impending collision of the Moon with Earth. This catastrophe forces them to make a choice that will determine the future development of human civilisation: '[i]t was necessary either to remake man's nature to suit another planet, or to modify conditions upon another planet to suit man's nature' (Stapledon, 1966, 246–47). They deem it impossible to alter any existing human for habitation of either Mars or Venus and decide that terraforming Mars would require them to bring water and air in quantities that would make such alteration near impossible. They therefore opt to terraform Venus rather than alter their own physical and mental characteristics to suit the planet. This decision implies that humanity will bring with it Earth's history when remaking other planets, because the tendencies that make up human nature guarantee a cyclical unfolding of Earth's cultural development on Venus, with all its historical oppressions and transient achievements. This is compounded by Stapledon's highly ambivalent treatment of the fifth men's genocide of the indigenous Venusians. Conflict between horror and guilt competes with the apparent necessity for self-preservation that initially drives the fifth men to terraform Venus. Anticipating Sparrow's concession that terraforming may be morally permissible if conducted to ensure humanity's survival, this episode raises questions regarding the problematic ethics of survival by portraying the fifth men's intellectual accommodation of the Venusian genocide: the narrator explains that, '[a]s for the murder of Venerian life, it was, indeed, terrible, but right. It had been committed without hate; indeed rather in love' (253).

Given the options of pantropy or the terraformation of Mars, this avowal appears a premature self-justification of imperialism. Commenting on Mark Sagoff's call for 'respect, regard, reverence, affection and love' as appropriate responses to nature, what Godlovitch calls an 'imperialism of respect,' the latter suggests that this 'can quickly degenerate into another nasty form of "respecting thine enemy" with its concomitant call to kill with kindness' (1994, 24, 26). Prior to this episode humanity had engaged in remaking human nature and, later, does so again. The third men realise that '[t]he present type of human being [...] was but a rough and incoherent natural product. It was time for man to take control of himself and remake himself upon a nobler pattern' (Stapledon, 1966, 150). The fifteenth men on Neptune also 'conceived, as an enduring racial purpose, the will to remake human nature upon an ampler scale' (280). Given the ubiquity of this theme, the genocide of the Venusians appears as a deeply ironic attempt to master both nature and other civilisations. When the fifth men successfully occupy Venus the narrator glosses two hundred million years in which 'all [the] main phases of man's life on earth were many times repeated on Venus with characteristic differences,' evidence of a repeated history that suggests the lingering presence of intellectual landscapes informing ideas of human nature and the significance of the spaces humans inhabit (258).

Deism and Teleology in Stapledon's Essays of Myth Creation

The eighteenth men are able to communicate telepathically with individuals existing throughout time. They also intuit, but are unable to communicate with, alien intelligences and their civilisations. These insights encourage the eighteenth men to speak of a 'spirit' or 'the Soul of All,' which awakens to conscious apprehension of itself and the universe (Stapledon, 1966, 309–10). In a passage that strongly anticipates Pierre Teilhard de Chardin's discussion of the noosphere in *The Phenomenon of Man* (1965, first published in 1955 but completed in 1938), biblical voices underlie the narrator's explanation that life aims towards consciousness:

> In the beginning there was great potency, but little form. And the spirit slept as the multitude of discrete primordial existents. Thenceforth there has been a long and fluctuating adventure towards harmonious complexity of form, and towards the awakening of the spirit into unity, knowledge, delight, and self-expression. And this is the goal of all living, that the cosmos may be known, and admired, and that it may be crowned with further beauties. (Stapledon, 1966, 307)

The narrator is fully aware that this belief is contingent. This image combines a pseudo-pantheistic, vitalist view of cosmological nature in which a 'spirit' pervades and constitutes all intelligent entities and a teleological view in which the development of consciousness, as an example of 'harmonious complexity of form,' is the goal of evolution. The narrator speculates on the possibility that 'the beautiful whole of things is the work of some mind; [...] [or] whether some mind admires it adequately as a whole of beauty.' The awakening of the spirit is likened to a vast music in which 'each individual factor is itself determinant, both of that which precedes and that which follows' (Stapledon, 1966, 312). Evolution and history are the two shaping factors of cosmological nature, which, in *Star Maker*, is the object of sublime aesthetic appreciation. The eighteenth men disrespect nature's otherness insofar as they deny nature's existence outside of this 'religious' or mystical belief. Nevertheless, the narrator, as one of the eighteenth men, admits that this view may not adequately account for cosmological nature.

In *Star Maker*, the disembodied group mind have long believed in and are searching for a great creator-destroyer-observer that they call the Star Maker. The existence of a being that creates universes for its own inscrutable purposes frames nature as teleological and so does not adhere to Lee's No-Teleology Thesis. This thesis rests on the distinction between teleology and teleonomy. Teleology offers explanations for an object's existence based on the assumption that it is organised so as to develop towards a predefined end point or goal. Because 'Earth (Nature) did not come into existence and/or continue to exist to serve human purposes,' it should not be viewed as teleologically oriented towards humanity. Instead, teleonomy is at work: 'there are teleomatic processes in abiotic Nature which simply follow physical laws, such as the law of gravity and the second law of thermodynamics,' and '[i]n biotic Nature [...] organisms display programmed behaviour, the programme being the product of natural selection' (Lee, 1994, 92). The No-Teleology Thesis claims that, although biotic and abiotic nature can serve purposes for humans, animals, and plants, neither exists solely for these purposes. In *Star Maker*, this thesis is compromised by a teleology oriented towards the figure of the Star Maker. *Last and First Men* appears to satisfy this condition, but the voice of a strong anthropic principle, in which the text's view of the cosmos suggests that its purpose is to provide the background for the full realisation of a human 'spirit,' goes some way to negating this thesis. Nevertheless, the portrayal of the cosmos in *Star Maker* supports the notion that this apparent teleology is a human interpretation, a landscape that only inadequately accounts for nature.

The Star Maker's relationship to the narrator is in many ways consistent with Hailwood's concept of nature's otherness, yet it is thoroughly overlaid with a deistic dimension. It is both the creator and the essence of cosmological nature; its disinterestedness and puissance mean that depictions of humankind's relationship to the Star Maker observe the Autonomy and Asymmetry Theses, if not the No-Teleology Thesis. Nevertheless, an argument could be advanced that nature here is in fact teleomatic: the sf representation of the radically alien re-contextualises the religious basis of the creator figure as godhead, recasting this image as a vector for nature's otherness as 'alien, inhuman, dark' (Stapledon, 2004, 96). The Star Maker does not quite capture the essence of a non-teleologic nature, although the narrator's inability to comprehend it opens up this potential: 'if he is nothing, if the stars and all else are not his creatures but self-subsistent, and if the adored spirit is but an exquisite creature of our minds' (99). As the narrator's experiences with the myriad inhabited worlds increase, he becomes more and more inclined to view the Star Maker teleomatically as 'unreasoning Creativity, at once blind and subtle, tender and cruel, caring only to spawn and spawn the infinite variety of beings, conceiving here and there among a thousand inanities a fragile loveliness' (135).

In a similar fashion, the third men in *Last and First Men* formulate a belief system based precisely on the conceptual surplus of nature's otherness, but they join this to a religious concept of 'Life as an all-pervading spirit, expressing itself in myriad diverse individuals' (Stapledon, 1966, 193–94). They engage in the torture of 'lower animals' through biological engineering as a form of worship, which offers a way to landscape nature that allows them to account for the environment's significance in human terms, even if their focus on pain disrespects the autonomy of other animals. They take as a blueprint for relationships between humankind and nature a view of evolution as a dialectic in which the environment functions as the background that shapes the development of life. The narrator notes that this 'backgrounding,' in which the environment's autonomy is denied, leads the third men to unsystematically emphasise physical pain as the chief aesthetic mode appropriate to respect for nature. Emphasis on pain reflects a Hobbesian conception of evolution as 'nature red in tooth and claw,' what the eighteenth men call 'Life's victory and defeat.' Elevating pain based on such intellectual landscapes leads to the uneven worship of the environment, which the narrator describes as a form of 'sadism' in itself (194). Nature thus conceived disrespects the No-Teleology Thesis because sentient life is positioned as dependent on humankind for its evolution and development.

This situation is grounded in the third men's disrespect of one aspect of nature's otherness: by landscaping sentient nature as 'lower' manifestations of their pseudo-pantheistic concept of life (sentience), and by complementing this belief with a philosophy that takes as a lesson from nature an emphasis on pain, they conclude that torturous biological manipulation leads to the closest possible apprehension of nature. As they believe themselves the highest expression of consciousness, they take it upon themselves to disrespect nature's autonomy and raise 'lower' animals to their status. The contradiction between respecting nature's otherness and landscaping nature by investing animals with the potential for transcendence recalls the theme of terraforming as the human duty to bring life to other planets. This episode parallels bodily modification and terraforming, demonstrating how the two endeavours grow out of similar beliefs about humanity's relationship to nature's otherness.

The narrator calls his vision of the Star Maker and the cosmos 'a fantastic reflex of itself, an echo, a symbol, a myth, a crazy dream, not wholly without significance,' which points to the impossibility of accounting for nature's otherness outside of necessarily partial human landscapes (Stapledon, 2004, 232). The Star Maker embodies the godlike power of creation; its ability to create universes of worlds parallels the aspiration behind the human endeavour to create new worlds via terraforming and, at greater levels of adaptation to the landscape, astrophysical engineering. The Star Maker can therefore be seen as the astrophysical engineer *par excellence*, although this would be to apply a particular landscape to a concept that retains a surplus of otherness.

When the narrator returns to Earth as the individual at the narrative's beginning, his experiences lead him to critique not only the limits of a science and philosophy that would pretend to a mastery of the world, but a complete knowledge that would manifest this disrespect of nature's otherness: 'Man's science was a mere mist of numbers; his philosophy but a fog of words. His very perception of this rocky grain and all its wonders was but a shifting and lying apparition' (Stapledon, 2004,

259). 'Man's science' offers a global sense of place that is completely constituted by human knowledge; it is a landscape that denies the presence of nature's otherness. The narrator contrasts this sense of place with 'the astronomical and hypercosmical immensity' and the smaller unit of community that his marriage represents, affirming the latter as the appropriate space for his limited human existence (Stapledon, 2004, 259). In contrast to Wells's release of experimental science from deductive intellectualism, Stapledon's narrator warns against a divorce between intelligence and 'spiritual sensibility': '[t]he mastery of physical force often produced a mania for power, and the dissection of society into two alien classes, the powerful and the enslaved' (246). The spiritual is emphasised as a necessary counterbalance to an exclusive reliance on a materialism that leads to a mastery of nature. The narrator of *Star Maker* does not deny the presence of nature's otherness; it is his spiritual experience of the cosmos and the Star Maker that lead him to affirm this vision of cosmological nature's alien otherness as an appropriate perspective for framing human relations.

The narrator's journey in *Star Maker* takes him through many worlds, and over its course he joins with other minds into a group consciousness that eventually extends into a cosmos-spanning mind. Yet, despite knowing 'the whole extent of space and time,' the I of the cosmical mind 'look[ed] about me with the same overpowering awe, the same abashed and tongue-tied worship as that which human travellers in the desert feel under the stars' (Stapledon, 2004, 212). When this cosmic mind confronts the Star Maker, the history of intellectual landscaping, in which notions of deities are used to explain the universe, helps conceptualise the awesome presence of this entity. The narrator characterises its vision of the Star Maker and creation as a 'fantasy' that its 'cosmical mind conceived' in order to comprehend just a small part of an experience that exceeds its ability to landscape (225):

> I, too, sought to capture the infinite spirit, the Star Maker, in an image spun by my finite though cosmical nature [...] this image, this symbol that my cosmical mind had conceived under the stress of inconceivable experience, broke and was transformed in the very act of my conceiving it, so inadequate was it to the actuality of the experience. (Stapledon, 2004, 227)

Awe is joined to love as the creation's response to the creator. The cosmic mind is rejected by the Star Maker, leading the narrator to realise that 'the creator, if he should love his creature, would be loving only a part of himself; but the creature, praising the creator, praises an infinity beyond himself' (Stapledon, 2004, 228). The Star Maker, whether conceived of as purely deistic or as an aspect of cosmological nature, is the object of the cosmic mind's love. Love flowing in the opposite direction would be narcissism. The narrator can only love the Star Maker as he conceives it, in an intellectually landscaped form that does not represent it accurately as subject or object.

Wells, Fearn and Stapledon engage with the debate about materialism versus vitalism that dominated ecological thinking during the interwar period. Unlike Wells, Stapledon's vision of the cosmos expresses elements of vitalism. Environmental narcissism, the image of nature as a reflection of humanity, is a theme that is often returned to in terraforming narratives such as Ray Bradbury's *The Martian Chronicles* (1958) and Kim Stanley Robinson's *Mars* trilogy (1996c [1992]; 1996b [1993]; 1996a). Terraforming disrespects nature's otherness insofar as it takes human landscapes to fully represent nature. In *The Shape of Things to Come*, humanity's physical mastery of nature is a direct expression of the urge to overcome its asymmetric relationship to nature's otherness, and in the novel's depiction of an increasingly technologised globalisation and humanity's future plans for terraforming Earth, it recapitulates this narcissistic landscaping. Fearn explores the opportunities for the technocratic organisation and management of Earth and other planets that science offers. *Star Maker* uses the radical otherness of cosmological nature to dethrone ideas of progress and of conquering the cosmos, which often masquerade as a desire for contact with otherness. Stapledon's ineffable Star Maker acts as a vector for nature's otherness and defies attempts to make the alien coherent with a particular anthropocentric paradigm. Examples of nature as cosmological, non-human and alien fundamentally undermine the protagonists' relationship to the cosmos, forcing them to re-evaluate their position within this new conception of nature. By realigning these relationships Stapledon does not emphasise humankind's connectedness to nature so much as he foregrounds the differing modes of otherness between humans and the rest of nature. Stapledon makes use of this dynamic of confrontation to explore and critique the assumptions underlying conceptions of human nature and cosmological nature's otherness. The narrator of *Star Maker* suggests that the narrative's attempt to reorient humankind's outlook through an extended engagement with philosophical speculation about cosmological nature may influence attitudes on Earth: 'perhaps the attempt to see our turbulent world against a background of stars may, after all, increase, not lessen, the significance of the present human crisis. It may also strengthen our charity toward one another' (Stapledon, 2004, 4).

Pre-1940s Proto-Gaian Living Worlds

A cluster of three short stories dealing with proto-Gaian living worlds was published between 1932 and 1934 in the American pulp market. Edmond Hamilton's 'The Earth-Brain' (1936 [1932]), Jack Williamson's 'Born of the Sun' (1934), and Laurence Manning's 'The Living Galaxy' (1934) are examples of what Mike Ashley calls 'cosmic sf,' which he traces to F. Orlin Tremaine's advocacy of the 'thought experiment' in 1933 (2000, 231). Ashley argues that the narrative thought experiment 'took space opera to its better extremes, considering not just the exploration of space but the nature of time, space and the universe' (231). The connection between Stapledon's use of cosmic scales to enlarge the scope of human perspectives and these proto-Gaian thought variants is suggestive of a critical element involved in imagining alternatives to conceptions of cosmological nature. Echoing Wells's account of the scientific organisation of the environment, Manning's *The Man Who Awoke* (1979 [1933]) reflects on the industrial overuse of natural resources and explores, using the language of ecological resource management, the shape of a future arboreal society. Manning's presentation of a planetary environment that has been actively forested and managed according to ecological principles positions this story as an early geoengineering text.

Two scientific romances familiar to Wells and Stapledon preceded this cluster of proto-Gaian pulp stories. M.P. Shiel's *The Purple Cloud* (2004 [1901]) portrays an early instance of a living world. Hamilton's 'The Earth-Brain' echoes elements of *The Purple Cloud*, establishing one of the early dialogues opened up by the motif. Arthur Conan Doyle was increasingly interested in the occult and spiritualism by the time 'When the World Screamed' (2000) was published in 1928, twenty-seven years after *The Purple Cloud*; that Shiel's, Doyle's and Hamilton's stories blend the supernatural and sf implies a connection between Gaia and spiritualism that Lovelock was unwilling to completely dissociate from the Gaia hypothesis. Jon Turney highlights Doyle's scientific romance as a potential precursor to Lovelock's Gaia hypothesis, and, though it is not certain Lovelock read it before 1965, Doyle's popularity and Lovelock's avowed love of sf suggest it is likely (2003, 77).

These stories situate their proto-Gaian living worlds within the bounds of the last of the *terrae incognitae*, at the Arctic in Shiel's and Hamilton's case and, in Williamson's and Doyle's, at the Earth's centre. 'When the World Screamed' frames the living Earth within the scientific paradigm of a hollow world, but, in contrast to Shiel's connection of Earth's processes to mythic themes, Doyle's use parodies scientific progress. The hollow world, already an outdated scientific hypothesis when *The Purple Cloud* was published, is reopened through the speculative element of the text to offer an alternative to the dominant view that 'the centre of the earth is liquid heat' (Doyle, 2000). In these scientific romances and the 1930s cluster of living world pulp stories, cosmic horror and an sf sense of wonder are the dominant responses to confrontations with threatening living worlds.

Proto-Gaian Scientific Romance: M.P. Shiel's *The Purple Cloud* and Sir Arthur Conan Doyle's 'When the World Screamed'

Shiel portrays the Earth as monstrous, thus associating horror with the little-known Arctic, a common *fin-de-siècle* motif of a landscape at the limits of scientific knowledge in 1901. *The Purple Cloud* echoes both Shelley's *Frankenstein* (1998 [1818]) and *The Last Man* (1996 [1826]): Adam Jeffson recounts his experiences as one of two survivors left on Earth after a volcanic cloud of gas exterminates humankind. This apocalyptic narrative develops the theme of isolation and offers a redemptive Adam and Eve myth set against the living planet motif, creating a space to focus consideration onto human responses to non-human nature. Jeffson encounters a lake at the pole, 'the old eternal inner secret of the Life of this Earth, which it was a most burning shame for a man to see [...] this fluid was the substance of a living creature.' He vaguely recalls 'a creature with many dull and anguished eyes' and 'the appalling nightmare and black abysm of sensations' this confrontation causes. Resonant with Stapledon's *Star Maker*, Jeffson relates his vision to 'fancy,' an 'impression, or dream, or notion,' and finally to 'madness' in an attempt to deny its existence (Shiel, 2004). This grotesque vision occurs after Jeffson crosses a boundary into a world aligned with darkness, nightmare and horror, thus establishing the structural theme of trespass. The creature itself is composed of fluid, which complements this trespass and potential breakdown of spatial boundaries with that of indeterminate monstrous form. The themes of the infinite ('eternal') and silence ('secret') are central to Jeffson's response to the living Earth.

Edmund Burke locates the origin of the sublime in objects that excite a sense of self-preservation and which therefore turn on feelings of pain or danger. He claims that 'terror is in all cases whatsoever, either openly or latently the ruling principle of the sublime' (1998, 102). Shiel firmly associates the creature with supernatural terror and intensifies this association by locating Jeffson's confrontation at the Arctic. The unfamiliarity of the landscape and the uncertainty of Jeffson's description tap into Burke's observation that obscurity enhances the terror of the sublime: '[w]hen we know the full extent of any danger, when we can accustom our eyes to it, a great deal of the apprehension vanishes' (102). This insight connects the search for comprehensive geographical knowledge, essentially a scientific and colonial quest of discovery and exploration, with the desire to encompass the obscure and to familiarise and therefore mitigate the terror and sublimity of the unknown.

Jeffson's encounter occurs in the supernatural context of a Manichean struggle between cosmic 'Powers' of light and darkness, a duality that is literalised by the Powers' internal struggle for Jeffson's soul. The catastrophe underscores humankind's fragility compared to a nature Jeffson sees as cruel and heedless. He describes a feminised Earth as 'dark and moody, sudden and ill-fated,' who 'rends her young like a cannibal lioness.' In another episode he uses this metaphor again when addressing nature, calling it a 'dark-minded Mother, with thy passionate cravings after the Infinite, thy regrets, and mighty griefs, and comatose sleeps, and sinister coming doom' (Shiel, 2004). George R. Stewart would later echo this theme in a biblical context in *Earth Abides* (1961 [1949]), which contains passages depicting Earth's autonomous existence after civilisation's destruction by an unknown plague. These intellectual landscapes associate Earth with the infinite, which Burke identifies as a major source of the sublime; Jeffson's ascription of infinity to the non-human world highlights his sense of an asymmetric relationship to Earth's unfathomable otherness. When Jeffson thinks that 'nothing could be more appallingly insecure than living on a planet,' he gives voice to Promethean fears that nature is not neutral but antagonistic to humanity. Conversely he also positively contrasts this intellectual landscape to the ice, arguing that 'the firm land is health and sanity, and dear to the life of man.' Shiel's use of the image highlights a negative, spiritual response to the immensities of Earth and the cosmos, but this relationship is ambivalent: he ultimately realises that the feminised Earth 'is old and wise [...] for great is the earth, and her Ages, but man "passeth away"' (Shiel, 2004).

Returning from the Arctic, Jeffson witnesses the effect of the still unknown natural catastrophe and experiences 'that abysmal desolation of loneliness, and sense of a hostile and malign universe bent upon eating me up' (Shiel, 2004). In his less lucid moments, Jeffson entertains the fear that his identity will be engulfed by the feminised Earth. Without a social context to provide points of recognition or self-definition, Jeffson finds it increasingly difficult to maintain his self-awareness: 'more and more the earth over-grows me, wooes me, assimilates me; so that I ask myself this question: "Must I not, in time, cease to be a man, and become a small earth, precisely her copy, extravagantly weird and fierce, half-demoniac, half-ferine, wholly mystic – morose and turbulent – fitful, and deranged, and sad – like her?"' (Shiel, 2004). What is feared here would later be sought after by others. This passage anticipates deep ecology's mystical and transcendent identification with nature, in which the subject's ego is subsumed. Jack G. Voller notes that this relationship was anticipated by 'the Romantic visionary mind,' which explored 'an identifying engagement with the numinous perceived to be informing nature' (1989, 138). Jeffson fears that nature will incorporate his identity, thus denying him autonomy and eventually leading to the elimination of his ego. Conversely, his forced isolation and his ascription of fitfulness, derangement and sadness to the Earth are instances of landscaping in which Jeffson expands his ego into nature. In either case, the theme of being 'one' with nature is implicit in this passage, and yet Jeffson recognises that this connection balances both similarity and difference. His relationship to nature retains an otherness that cannot be overcome by knowledge, yet identification can be predicated on shared origins:

> [h]er [Earth's] method of forming coal, geysers and hot sulphur-springs, and the jewels, and the atols and coral reefs; the metamorphic rocks of sedimentary origin, like gneiss, the plutonic and volcanic rocks, rocks of fusion, and the unstratified masses which constitute the basis of the crust; and harvests, the burning flame of flowers, and the passage from the vegetable to the animal: I do not know them, but they are of her, and they are like me, molten in the same furnace of her fiery heart. (Shiel, 2004)

This view of Earth's natural processes breaks down the radical separation of the human/nature duality by recognising humankind's shared origins with diverse non-human others. It erects boundaries by establishing distinctions between humankind and nature as much as it offers identification. Jeffson's landscapes keep nature's otherness in view even as they attempt to reduce difference. They negotiate between natural processes and a mysticism that is productive of Promethean fear and the sublime.

While Shiel focuses on the individual's sense of cosmic horror towards the immensities of Earth and the cosmos, he nevertheless assigns to the social sphere responsibility for the projection of intellectual landscapes derived from human experience:

> Man's notion of a Heaven, a Paradise, reserved for the spirits of the good, clearly arose from impressions which the earth made upon his mind: for no Paradise can be fairer than this; just as his notion

> of a Hell arose from the squalid mess into which his own foolish habits of thought and action turned this Paradise. (Shiel, 2004)

An interest in social worlds accompanies portrayals of an individual's experience of cosmic horror. Jeffson's claim that paradisal landscapes are based on experience of non-human nature while their converse is a consequence of culture assigns accountability for the creation of nature's fear-inspiring asymmetry not only to the human management of Earth's resources, but more generally to humankind's politico-cultural attitudes. Jeffson landscapes the planet as a terrestrial ship, while a literal ship also provides transport for the Arctic expedition and Jeffson's later international travel. His experiences during these solitary years lead him to orient himself towards civilisation's ruin and nature as the only other sources of interaction. He metaphorically extends his phenomenal experience of the world to Earth's processes; by relating the eruption of the purple cloud in language that ties it to the beating of waves against his ship, he establishes a relationship of microcosm–macrocosm between his journey and 'this planetary ship of earth,' using this model to understand the catastrophe as 'a wave rather which she [the Earth] had reserved, and has spouted, from her own un-motherly entrails...' (Shiel, 2004). In this context the catastrophe itself can be read as a punishment for civilisation's transgressions. The image of the Earth as a ship sailing through space contributes another dimension to Jeffson's anxiety towards nature by foregrounding the planet's fragility. This metaphor, like Wells's reference to an abundant Earth opened up by geoengineering, echoes Henry George's description of Earth as a well-provisioned ship. George uses this image to highlight the human exploitation made possible by an elite's possession of a planet's abundant resources, thus grounding his arguably justified faith in Earth's material ability to sustain the human population in 1879.

In Arthur Conan Doyle's 'When the World Screamed,' George Edward Challenger, the distinguished if intolerant professor of *The Lost World*, sets out to prove his hypothesis that Earth is an organism whose skin is the eight-mile crust of the planet's surface. Challenger contracts the narrator, Peerless Jones, to complete the final stage of a drilling project aimed at piercing the flesh of the uncovered organism. Challenger considers this project 'one of the greatest experiments – I may even say the greatest experiment – in the history of the world' (Doyle, 2000), and explains that '["]the world upon which we live is itself a living organism, endowed, as I believe, with a circulation, a respiration, and a nervous system of its own." Clearly the man was a lunatic' (Doyle, 2000). Challenger's thesis anticipates the concept of planetary homeostasis, while Jones's reaction prefigures the resistance Lovelock faced from the scientific community upon first proposing the Gaia hypothesis. Lovelock retrospectively claimed the Scottish scientist James Hutton as one of the precursors of the Gaia hypothesis, who argued in 1785 that 'the Earth was like an animal and that its proper study should be by physiology' (Lovelock, 1995, 9). The French physiologist Claude Bernard first recognised organisms' self-regulating properties as the 'wisdom of the body,' which Walter Cannon referred to in the 1930s when he coined 'homeostasis' (Lovelock, 1995, 18–19). Lovelock calls the study of Earth as Gaia geophysiology, which draws together this combination of the physiological and geophysical sciences and echoes what is made literal in this story: traditional literary images of a personified or zoomorphic Earth that act as intellectual landscapes prefiguring a colonial approach to physical space.

Challenger's ostensible motivation is the search for scientific knowledge: he points out that '[t]o know once for all what we are, why we are, where we are, is that not in itself the greatest of all human aspirations?' Behind this lies a narrative of conquest that informs his language: 'I propose to let the earth know that there is at least one person, George Edward Challenger, who calls for attention – who, indeed, insists upon attention' (Doyle, 2000). Having dominated the academic sphere and amassed a fortune, his ambition escalates, but his declaration suggests that underlying this disinterested scientific inquiry is an infantile desire for recognition and attention. His desire, as two of the characters point out, is coupled to a general lack of care and respect and is overlaid with personal aggrandisement, qualities that contribute both to his notoriety and to the appeal of his entrepreneurial individualism.

Edward Malone calls Challenger 'a primitive cave-man in a lounge suit,' but clearly admires his achievements, also calling him 'the greatest brain in Europe, with a driving force behind it that can turn all his dreams into facts.' The scientific pursuit of knowledge, joined to a disregard of its effects, is characterised as a fundamental yet primitive trait of human nature. Despite many lawsuits, Challenger continues to show little respect for the environment. In an episode that exhibits an early interest in issues of environmental preservation, Malone reports how Challenger '[s]aid [the machinery] was one-tenth of an inch out of estimate, so he simply chucked it by the wayside.' Expressive of fears that the British countryside would be despoiled at the whim of entrepreneurs and industrialists, the narrator explicitly aligns the mistreatment of social and natural worlds by the wealthy and pioneering scientist. He explains that '[a]n audience after one of Challenger's harangues usually felt as if, like the earth, its protective epidermis had been pierced and its nerves laid bare' (Doyle, 2000). Challenger's achievement in penetrating Earth's eight-mile crust is emblematic of his capacity for confrontation and mastery.

Jones, an expert in artesian borings, enacts the relationship of dominance Challenger poses to the Earth. Challenger internalises the gap between himself and others, describing Jones in unflattering terms as a mechanical instrument to be directed by his intelligence and will. He naturalises this relationship when he claims that '[a] certain analogy runs through all nature': '[y]ou, sir, represent the mosquito. Your Artesian borer takes the place of the stinging proboscis. The brain has done its work. Exit the thinker. Enter the mechanical one, the peerless one, with his rod of metal. Do I make myself clear?' (Doyle, 2000). This insect analogy anticipates Hamilton's and Williamson's use of the trope and is reiterated in many later living world narratives, standing as a model of humankind's dependence on Earth. This asymmetry is connected to another prevalent symbol, the gender-coded drill that stands opposed to the feminised Earth. Upon seeing the Earth's flesh, Jones exclaims, 'Good Lord! [...] And am I to plunge a harpoon into that beast!' thus paralleling Challenger's hubristic domination of the landscape and Ahab's attempt to dominate the sea and its eponymous denizen in *Moby-Dick*. The narrator's closing words, '[i]t has been the common ambition of mankind to set the whole world talking. To set the whole world screaming was the privilege of Challenger alone,' locate Challenger at the forefront of civilisation's progress, an achievement made possible by an ethically vacuous science and technology that does not in itself represent any progress in human nature, but rather represents an extension of basic human responses to nature. Jones refers to the living planet as 'Mother Earth' when he credits Challenger with an ironised panegyric at the successful conclusion of the project – 'Challenger the super scientist, Challenger the arch-pioneer, Challenger the first man of all men whom Mother Earth had been compelled to recognize' – and reports that 'nowhere did the injured planet emit such a howl as at the actual point of penetration, but she showed that she was indeed one entity by her conduct elsewhere' (Doyle, 2000). The image constructed is of a scientist raping the world, an action supported by a self-centred application of technology.

Jones witnesses a scene that indexes an alternative mode of relating to nature's otherness. A system of elevators from the Earth's surface to its unshielded flesh operate as a metaphorical time machine through geologic time. The strata revealed by the borings are a sign of Earth's age, and Jones's reaction to this signifier of the infinite is one of wonder: '[t]he archaic rocks varied wonderfully in colour, and I can never forget one broad belt of rose-coloured felspar, which shone with an unearthly beauty before our powerful lamps' (Doyle, 2000). 'Unearthly' is a strange adjective to attach to the earth, one which paradoxically constructs our native planet as alien and reveals the extent of the limits to intellectual landscapes when confronted with the Earth's strata, a geography where temporal estrangement increases the further down you go (down is back in time). Jones's sense of awe and wonder towards the alien beauty of Earth's geology and age offers an alternative to Challenger's antagonistic attitude to nature's otherness. Earth's familiarity is a consequence of interactions between humankind and nature that elide its alien otherness. Despite Fogg's complaint that 'terraforming' describes the task of adapting planets to resemble Earth and so cannot be applied to it, the phrase 'an unearthly beauty' calls attention to the longstanding human project of adaptation, in which the Earth has been landscaped both intellectually and physically in order to shape it to humanity's needs and desires.

The Pulp Sf Proto-Gaian Cluster

Edmond Hamilton's 'The Earth-Brain' describes the otherness encountered by explorers at the far regions of the Earth. Landon recounts his expedition to the Arctic, where he discovers a mountain that literally houses the Earth's brain. The expedition dismisses the accompanying Sherpas' warning of a mythic prohibition on approaching the mountain, but one member entertains the possibility of its truth:

> Why couldn't earth be a living organism instead of just a mass of inanimate matter? It seems an inanimate mass to us, it is true, but so must a human being seem an inanimate mass to the microbes that live on and in that being. Earth might be a living organism, all the planets might be organisms, of scale and nature so different from us that we mites who swarm upon it cannot even comprehend it. And if it is living it could possess consciousness and intelligence, perhaps intelligence operating on planes and for ends entirely alien to us. (Hamilton, 1936, 144)

Both Lovelock and his collaborator Lynn Margulis employ the same analogies concerning the relationship between macrocosm and microcosm, between Earth's Gaian system and its constituent parts. Speculative questioning extends the living world and the scope of *terrae incognitae* throughout the solar system. The assertion that the human cognitive faculty cannot encompass this larger system reinforces the

Asymmetry Thesis by fuelling a sense of cosmic horror, itself a form of Promethean fear. Crucially, non-human nature is not inert matter explicable in mechanistic terms, but an organic and autonomous creature in its own right. Ernest J. Yanarella identifies such challenges to the rational-mechanistic landscape as central to later treatments of the Gaia hypothesis (2001, 250–51).

Morris warns the reader of humankind's endemic lack of knowledge in contrast to a worldview that promises coherent and complete knowledge of nature: 'in your unbelief remember this – that of all things in the universe we men know least really of this earth we live upon' (Hamilton, 1936, 137). The expedition discovers truth behind the myth when they encounter 'a giant ovoid of light or force that towered there at the cone-cavern's centre' (157–58). This meeting echoes the confusion attendant on confronting the supernatural entity in *The Purple Cloud*. The ovoid inspects and subdues the interlopers with 'senses having nothing to do with any senses we knew but operating on planes entirely different'; Landon relates how '[t]he impact of that will was tangible, overwhelming. It seemed partly to replace, to usurp, my own will and mind' (Hamilton, 1936, 160, 163). Subversion grants a brief psychological connection to the alien intelligence, and through the resulting human/non-human duality of mind he is granted an awe-inspiring vision of the universe: '[m]y great body was racing at awful speed through vast leagues of infinite space! Far off across those immensities of space I was aware of other living earths, other planets, some larger and some smaller than I, but each living in the same vast way as I lived, each with its own great brain!' (Hamilton, 1936, 164–65). Communication with a planetary alien intelligence appears in embryonic form where contact involves an asymmetric non-human dominance of the channels of communication. Human consciousness approaches indistinguishability from the Earth-Brain, leading to a model of the transpersonal self in which the overcoming of one's ego is inverted to become an undermining of that ego. This scene taps into the sense of cosmic horror through the image of a universe populated by unfathomable living planets, symbols of cosmological nature's otherness that confound the expedition's search for scientific knowledge. The explorers attempt to gain a measure of control over nature by occupying it through exploration and subsumption into their scientific schema, thus annexing it to the human sphere and reducing its otherness to an identity.

The expedition's ethnocentrism is challenged by a contrast between their worldview and 'traditional' knowledge from other cultures, thus undermining the stability of their understanding of the universe and deflating their faith in reason and science. A critique of colonial tendencies informs this aspect of the text, as the Sherpas take on the role of representatives of a primitive society unclouded by faith in the comprehensiveness of scientific knowledge. This contrast implicitly extends the text's scope to include other ideologies underpinning notions of Western civilisation and progress. The Sherpas are guides and harbingers of an ancient and dangerous knowledge, warning of 'the forbidden mountain at the earth's top – shunned by all our race!' (Hamilton, 1936, 142). They are also ciphers whose role is to provide a foil for notions of a Western science allied to a dogmatic view of the non-human. When the Earth-Brain kills the other two members of the expedition, Landon fires at it in self-defence. His action is described as a sin against nature, a transgression of the mythic prohibition spoken of by the Sherpas: '[c]olossal anger emanated from it at the same moment like a wave of destroying force, and as that cosmic wrath swept through me I knew that I had committed blackest sin against the universe in daring to attack the brain of the living earth-body upon which dwelt I and all my tiny race!' (Hamilton, 1936, 170). The Earth-Brain exacts retribution for Landon's violent reaction against the asymmetrical relationship between cosmological nature and humankind. The Earth-Brain's earthquake-inducing pursuit of Landon across the globe suggests the theme of humanity's regulation as part of a system and assigns that responsibility to the Earth. The living planet operates in this capacity as a check to human domination, with tremors operating as a sublime symbol of Earth's reckoning. Morris implies as much when he warns that 'we who consider ourselves masters of all are not but a race of microscopic parasites dwelling upon the vast and strangely living body of that Earth-Brain' (Hamilton, 1936, 181).

Jack Williamson's 'Born of the Sun' (1934) negotiates the boundary between cosmic horror and the sf technological sublime. Earth and the other planetary bodies are eggs, their parent the sun, which 'expands and contracts in the rhythm of the sun-spot cycle, with a beat like the pulse of a living thing' (Williamson, 1934, 16). The Moon's hatching prefigures Earth's fate and relates dread and majesty when the narrative characterises the living world as 'more than anything else like the eldritch, gorgeous streamers of the Sun's corona' (25), thus tapping into the language of weird fiction through the adjunct 'eldritch,' often associated with Lovecraft and meaning '[w]eird, ghostly, unnatural, frightful, hideous' ('Eldritch, adj.'). Further description of '[a] body, both horrible and beautiful' and reference to '[t]he shadows it cast, inky-black, green-fringed, [which] were uncanny – dreadful,' show an ambivalence that accompanies revelation of nature's alien otherness and subverts dominant understandings of Earth (Williamson, 1934, 25).

Williamson connects the sublimity of the living world to beauty, two concepts that Burke considers separately as centring respectively on pain and pleasure. Beauty for Burke is a social quality that attracts humans into an engagement with the other. Unlike the sublime, it is founded on the submission of the object to the subject: 'we submit to what we admire, but we love what submits to us; in one case we are forced, in the other we are flattered into compliance' (Burke, 1998, 147). The tension between the sublime and beauty is an important cognitive structure for framing human relationships to the non-human. Ultimately, it suggests that beauty as an aesthetic response to nature's otherness is complicit in domination of nature and so cannot offer aesthetic grounds for ecologically sound relationships to the non-human. Yet Patrick D. Murphy highlights the androcentrism implicit in the notion that triumphing over the fear, awe or horror engendered by sublime landscapes is one source of the pleasure derived from the experience of a dangerous nature (Murphy, 2012, 82). This sense of triumph also undermines the Burkean sublime as an appropriate aesthetic response that would nurture ecologically sound relationships to nature. Murphy explores an ecofeminist revisioning of the sublime and points out that while the '[r]ecognition of the physical limitations could also lead to an awareness of human interconnectedness and interdependency with other human beings and other living entities on the planet,' this dependency is not emphasised 'in most literary representations of the sublime' (83). In many of the works examined in this chapter, humankind's physical limitations are emphasised in such a way as to highlight humankind's fundamental dependency on non-human nature.

Barron Kane is another explorer figure, who discovers and brings knowledge of the Earth's nature to his nephew Foster, an inventor and the story's protagonist. Kane has infiltrated a mysterious sect where, echoing Hamilton's use of Sherpas as homogenised foils for Western (rational scientific) landscapes, he discovers 'that oriental insight had seen the truth hidden from our dogmatic western minds' (Williamson, 1934, 16). Acting against the sect's fatalistic belief that humanity should resign itself to extinction during Earth's catastrophic birth, Foster and his uncle, accompanied by Foster's fiancée, June, and a selection of colonists, hasten to construct a functioning spaceship to escape Earth's destruction. Social responses to the catastrophe are subordinated to Foster as heroic individual. Viewing the Moon's destruction, Foster and June affirm the importance of the smallest unit of community as a response to individual insignificance:

> 'It was lovely – and horrible –'

> [...]
>
> 'Our world must go – that way, dear –' he breathed; and her shivery tiny whisper finished: 'But we have – each other –' (Williamson, 1934, 25)

Community is conveyed by romantic cliché, but Foster's status and the masculinist heroism of the text emphasise his individualism and hubris. He is elevated from the group in a version of the transcendent self when, '[i]n a moment of crystal vision,' he sees himself 'not as one man fighting for his own life, but the champion of humanity, battling for ultimate survival' (Williamson, 1934, 20). Barron Kane supports Foster's vision when he says '[i]t's up to you, now, to save the seed of mankind,' and he evokes the Adam and Eve myth also utilised in *The Purple Cloud*, but transplanted to the geography of space: '[t]he children of Foster and June will conquer space, to the farthermost one of you!' (38). The obligation to remain independent of the planets is incompatible with this colonial urge to conquer space. Foster's attempt to combat both nature and the mobs incited by the sect is a struggle between submitting to the other and forcing the other into submission.

Invention gives humanity a way to free itself from its dependency on nature. The spaceship, christened the *Planet*, 'can sail on forever, Barron. It's a little world, itself, independent of the Sun' (Williamson, 1934, 22). The insect metaphor is revisited in conjunction with this theme when Foster realises that, because of his anti-gravity invention, 'Men will now be small parasites no longer, to be crushed like vermin by any chance tremor of the beast that bears them' (38). The machine grants an affective response opposed to the helpless awe they experience when seeing the moon crack: '[a] kind of lofty elation' and 'a sense of triumphant power that lifted him far above any human concern' (33). This allows Foster to transcend his ego and attain 'the supreme tranquility of a god [...] It was sublime, awful Nirvana. He had forgotten even June' (33). This alternative to cosmic horror, which draws from the Kantian sublime's intellectual transcendence over a dynamic nature, means that humankind can now begin to occupy a position of dominance in relation to nature. Foster's recognition of humanity's dependence on the planets, each with an existence independent of humankind, implies a moral dimension suggestive of obligation towards, and an opportunity to achieve hyperseparation from, nature: '[y]ou're alive, all of you. We owe our lives to you – we've been parasites on your kind. But we aren't any longer. We're beginning all over again, on our own' (38).

Meeting this obligation coincides with the promise of a new, utopian society represented by the crew of the *Planet*, '[s]ix hundred picked men, representing every race and every craft and every creed, with their wives and children. Two thousand all told – and the very cream of humanity' (Williamson, 1934, 22). There are problems with this basis for a new humanity that are especially evident when its members are contrasted to the orientalised Eastern sect opposing them, members of which are described as 'yellow-visaged demons armed with the weapons of a secret science' (28). The sentence's syntax invests only the six hundred men, not their wives (the only women aboard) or children, with the power to represent humanity's diversity. In addition, the selection process is suspiciously elided. The sect's manipulation of human fears during the tremors presaging Earth's traumatic cracking allows them to muster an army to confront and kill the colonists. Their deaths further elevate Foster and June's personal relationship.

Laurence Manning's 'The Living Galaxy' (1934) prefigures Stapledon's extension of the living world trope to galactic scales, offering a series of macrocosmic levels for the colonial extension of Earthbound life throughout cosmological nature. The central story is narrated as a history lesson for a child of the far future and relates with some uncertainty an explorer's encounter with a potentially living galaxy, along with contextualising information about a now ancient and mysterious Earth. An initial frame narrator contemporaneous with the implied reader begins with an apology and a request: '[i]t is impossible for me, as author, to write their story so that it is complete in itself; I must ask you, as reader, to lend a hand to the work' (Manning, 1934, 437). This device calls on the reader to imaginatively assume the identity of a child at a history lesson in the far future while engaging a critical stance towards the events related during the second frame narrative of the history lesson. The narrative's discussion of colonial expansion throughout the galaxy, which Foster anticipates with wonder in 'Born of the Sun,' is destabilised by positioning the second frame in the far future. A space is thus introduced to reflect on the past, contributing to the irony directed to notions of technological progress.

Manning's awareness of ecology, demonstrated a year earlier in 'The Man Who Awoke,' informs principles central to the technical aspects of terraforming in 'The Living Galaxy.' The narrator observes that Earth 'possessed by nature a climate and an atmosphere suitable to human existence without any artificial aide' and 'was deserted by thousands of explorers who settled down on the five remaining planets of the solar system. These were not habitable without artificial air and heat' (Manning, 1934, 438). Manning's interest in planetary environments and his knowledge of ecological issues stem from his interest in space colonisation and his involvement in pioneering early space rocketry. He was one of the founding members of the American Interplanetary Society, which changed its name to the American Rocket Society and merged, in 1963, with the American Institute of Aeronautics and Astronautics (AIAA). As the editor of the American Interplanetary Society's journal *Astronautics*, Manning was well informed with regard to the technical aspects of spaceflight and the maintenance of contained environments. In the mid-1940s Manning retired from the organisation, an event that coincided with a shift towards the professionalisation and legitimation of rocket science as a field for scientific and technical enquiry. Manning also wrote *The How and Why of Better Gardening* (1951), from which could be inferred a link between spaceflight and garden landscapes.

In 'The Living Galaxy,' early terraforming themes are connected to the colonisation of space, which, as in Williamson's story, is reliant on technology for realisation. Colonisation is seen as a case of 'steady, peaceful expansion' made possible by 'two great inventions' and cosmological geography: the release of atomic power and atomic synthesis (anticipating Williamson's CT stories), and the abundant space and resources available throughout the galaxy (Manning, 1934, 438). This latter idea alludes to histories of the colonial acquisition of resources and territories, a relation that is strengthened by the pedagogic narrative frame. However, the initial narrative device, by highlighting the textuality and uncertainty of the fictional historical events, engages the implied reader in an ironic unmasking of the official story, thus calling into question notions of colonial expansion.

Bzonn's encounter with what he believes is 'a gigantic creature rooting dangerously with a tentacle among the stars that housed the human race' shows that despite this ideology of peaceful expansion, a meeting with nature's alien otherness provokes a violent reaction (Manning, 1934, 442). The narrator, whose authority is already compromised by his temporal distance from these events, asserts that '[i]t must be borne in mind that Bzonn felt no doubt that the star-mass composed a living intelligent creature' (442). The teacher's professional scepticism leads him to consider various opposing theories, but the structure of the narrative manipulates the reader into support for Bzonn's theory that 'the protuberance was a creature of life in some form which utilized solar systems after the fashion of atoms' (441). Bzonn's destruction of the galaxy in perceived self-defence means that certainty can never be attained, much to the implied detriment of humanity.

The story's 'Afterword' sees the return of the initial frame narrator but is narrated in second person, towards an implied reader who retains their imaginative assumption of the schoolchild's identity. This device constructs a complex set of relations between colonial expansion and a child's curiosity and wonder towards grand historical events, which introduces a subtext that is critical of expansionist ideology. The narrator conflates the supposed ideological position of the implied reader and this member of the far future when describing how '[y]ou are one of those who cannot wait for the next day to bring what it will – you must peer into the next chapter, driven by curiosity. For long hours you sit there over the book and I would give anything to know what you read there!' (Manning, 1934, 497). The frame narrator, implied reader and child of the far future all have a stake in the knowledge and wonder of the future, and yet this intergalactic imperialism is undercut, thus placing this story in opposition to the faith in scientific rationalism and technology evident in Williamson's 'Born of the Sun.'

This discussion highlights a link between early terraforming, ecological catastrophe and proto-Gaian narratives, namely an asymmetry that Yanarella identifies as a 'profound indifference' towards the fate of individual species, a characteristic that he sees as implicit in the Gaia hypothesis (2001, 227). Lovecraft describes the fear of the unknown as productive of cosmic horror, an experience that, like Promethean fear, fundamentally encodes an awareness of the asymmetric relationship between the human and non-human (2009b [1927]). Short stories such as 'The Colour Out of Space' (2009c, [1927]), 'The Dreams in the Witch House' (2009d, [1933]) and 'At the Mountains of Madness' (2009a, [1936]) present cosmic horror in relation to aspects of a previously unsuspected cosmological nature. The echoes of Shiel's text in Hamilton's treatment of the living world and their use of cosmic horror further support the case for the dialogic exploration of proto-Gaian themes in scientific romance and early pulp sf. Proto-Gaian themes are connected to journeys of discovery, a structure that opposes known and unknown spaces as an emblem for scientific discovery, itself a method of landscaping nature's otherness. Williamson's text offers the technological sublime as a counter to the sense of human insignificance in the face of an awe-inspiring cosmos, whereas the discovery of the proto-Gaian entity in other stories challenges the capacity for scientific enquiry and colonial projects of acquisition and mastery to erase nature's otherness.

The Decline of the Living World Motif in 1950s American Pulp Sf

As has been shown, terraforming and living world stories offer fertile spaces for enviro-ethical reflection. The relationship between nature's otherness and the theme of interconnectedness, a significant aspect of the Gaian texts influenced by Lovelock's hypothesis, has already been anticipated by episodes in which the line between human and non-human nature is blurred, as in *The Purple Cloud* and 'The Earth-Brain.' Few living world stories appear in the 1940s–1950s, a period overlapping with the development of the first phase of narratives that make terraforming their focus. The living world narrative sporadically resurfaces during the postwar period, with the result that the motif becomes situated ever-more firmly in the American pulp sf tradition.

Murray Leinster's 'The Lonely Planet' (1949) is one of few living world short stories published after *Star Maker* and before the expansion of the American environmental movement in the 1960s. The living world in this story is set against an interplanetary background, geography already touched on in Stapledon's and Manning's work. Through an alien planetary consciousness, Leinster continues his exploration of human–alien contact famously handled in his short story 'First Contact' (2004 [1945]). The living world is relocated from the planet's core to its surface. Christened Alyx by the first expedition to discover the planet, the creature is a single living organism that develops consciousness after exposure to humankind. At first its compliance with men's wishes makes it 'a living, self-supporting robot, an abject servant to any creature with purpose it encounters,' and it is exploited for its labour in mining the valuable rotenite in its crust (Leinster, 1949, 83). This focus on the exploitation of natural resources aligns this story with Doyle's 'When the World Screamed' and the instrumental relationship to nature signified by the motif of the drill.

When the Alyx Corporation discovers that the planet's scientific and technological progress has outstripped humanity's, their economic interest in Alyx leads them to the view that '[t]he idea of a greater-than-human intelligence [...] [is] frightening. If it became known, the results would be deplorable' (Leinster, 1949, 85). They decide that 'Alyx had to be killed,' because '[i]t was wiser than men. It could do things men could not do. To be sure, it had served mankind for five hundred years' (90). Alyx is capable of developing technology incomprehensible to humanity, but the humans' fear leads them to reject opportunities to establish a mutually beneficial relationship. Alyx offers to create a utopia for humankind on its surface, yet, except for a few descended from the explorer who originally identified Alyx as sentient, it is refused. Recalling the orientalised cults in Williamson's 'Born of the Sun,' the narrator records how '[c]ults, too, sprang up to point out severally that Alyx was the soul-mother of the universe and must be worshiped; that it was the incarnation of the spirit of evil and must be defied; that it was the predestined destroyer of mankind and must not be resisted' (Leinster, 1949, 94). These responses are indicative of attempts to explain in religious terms threatening cosmic events, thus connecting the societal confusion caused by a confrontation with nature's otherness to the dialogic aspect of sf, here represented as a thematic element of the text. Several groups adhering to diverging positions speak for Alyx, thus demonstrating attempts to incorporate the living planet into humankind's intellectual landscapes to downplay its radical alien otherness. In contrast to the dangerous relationship that the living world poses to humankind in many of the stories discussed above, Leinster portrays the living world in a manner resonant with Fearn's depiction of the alien scientists in 'Earth's Mausoleum.' The benefits that Alyx offers are refused because of limitations in humankind's ability to accept an asymmetry between itself and nature, which generates Promethean fear and leads to its attempt to gain mastery over the living world through force.

These works offer textual sites for philosophical reflection and are entangled with other treatments and themes as part of the megatextual dialogue of sf. Nature's otherness, a concept designating the relationship of non-human nature to human nature and culture, is understood in cosmological terms in the scientific romances, early cosmic pulp sf and postwar pulp sf dealing with terraforming, geoengineering and living worlds. The thematic prevalence of resource extraction in these proto-Gaian texts links them to notions of terragouging, the terraformation of a planet 'to facilitate extraction of raw materials for earthly consumption' (Murphy, 2001, 270). The drill signifies humankind's exploitative relationship to nature, functioning as a metonymy for a civilisation that homogenises and landscapes it in terms that oscillate between monstrosity and motherliness. These proto-Gaian narratives disrespect nature's otherness, but include interstices through which recognition of radical otherness enters the text as one voice among others.

Lee's Asymmetry, Autonomy and No-Teleology Theses help identify the fundamental environmental attitudes that structure human relations to nature's otherness. Promethean fear, grounded in the axiomatic asymmetry between humankind and nature, is evident in the scientific romances of Shiel and Wells and in the cosmic sf of Hamilton,

Williamson, Manning and Leinster. Such responses are imagined as leading towards an urge to direct humankind's environments and history via technocratic societal management, as in *The Shape of Things to Come*, or through various technological fixes such as the contained spaceship/habitat in 'Born of the Sun.' When the Asymmetry Thesis is undermined in cosmic sf, Promethean fear is overturned and technology appears to elevate humanity above nature, allowing it to exploit and control the cosmos. Manning's 'The Living Galaxy' ironises this development and, along with Stapledon's *Star Maker*, challenges the colonial anthropomorphism of the war on nature theme that informs terraforming and living world stories by offering vectors for recognising nature's otherness.

This chapter has shown how terraforming and proto-Gaian themes were used to explore civilisation's disrespect of nature's otherness in works published during the interwar period. The next chapter explores how terraforming was framed in terms of the American pastoral in the American pulp sf of the 1950s–1960s.

2: The American Pastoral and the Conquest of Space

The first book-length narratives that dealt with terraforming as part of their foreground or as their organising motif were published in the 1950s: Ray Bradbury's *The Martian Chronicles* (1958 [1950]) collected his 1940s short stories into a single volume that portrayed the colonisation and terraformation of Mars; Robert Heinlein's *Farmer in the Sky* (1967 [1950]), set on Earth and Ganymede, was the first novel to dedicate its narrative to terraforming; and Arthur C. Clarke's *The Sands of Mars* (1976 [1951]) looks with irony at the pastoral romanticism that informs the language of stories of colonisation. These texts are offset by another collection of texts that offered a polar response to the utopianism depicted in these early 1950s stories. They used the motif of planetary adaptation to explore the inequalities and excesses of humankind's social and political relations. Such dystopian narratives include Frederik Pohl and C.M. Kornbluth's *The Space Merchants* (1974 [1953], first serialised in 1952 as 'Gravy Planet'), Walter M. Miller's 'Crucifixus Etiam' (1973 [1953], first published as 'The Sower Does Not Reap'), and Poul Anderson's 'The Big Rain' (2001 [1954]) and *The Snows of Ganymede* (1958 [1955]).

In the late 1950s to early 1960s the narratives of the first terraforming boom began to shift ever-more emphatically towards the terraformation of planets inhabited by alien species. The shifting foci of these stories index a corresponding shift of concern in the sf of this period and a transformation of how terraforming was used to comment on society. The worlds imagined in these works helped shape a tradition of consensus about the possible futures that might develop from the trends of the 1950s. Central to this image of time is its construction through a dialogue between texts, one whose nature became increasingly sophisticated as it was deployed and re-deployed to explore social, technological, political and philosophical concerns.

These terraforming stories often deployed pastoral motifs and structures to engage in socio-political enquiry. In his essay on novelistic chronotopes, Bakhtin classifies the pastoral as a subset of the idyll, which he isolates in four 'pure types: the love idyll (whose basic form is the pastoral); the idyll with a focus on agricultural labor; the idyll dealing with craft-work; and the family idyll,' but he acknowledges that there are 'mixed types [that] are extremely widespread, in which one or another aspect predominates (love, labor, or family)' (2002, 224). *Farmer in the Sky,* like many other stories, connects terraforming to agricultural labour and traditions of American agrarianism. Elements of the family idyll are also mobilised to explore ideas of community and to parallel the maturation of the protagonist with his efforts to make a new home by transforming Ganymede.

Apart from these distinctions of type, Bakhtin also notes stylistic gradations within the same type: 'distinctions in character and degree in the metaphorical treatment of individual motifs (for example, natural phenomena) as they are incorporated into the totality of the idyll, that is, differences in the extent to which purely realistic or metaphorical links predominate, differences in the degree and nature of the sublimation and so forth' (Bakhtin, 2002, 224). Bakhtin's analysis of these pastoral types parallels Andy Sawyer's observation that the pastoral and sf are dynamics rather than genres whose interplay within a given text helps generate its meaning (2006, 396). The combination and variation of elements within and between types, all of which modulate the idyllic chronotope in one of its 'pure' forms, suggest a dialogic process at work. Given these variations, Bakhtin argues for a core of common features shared by all examples of the idyll that are made coherent by 'the special relationship that time has to space' in these narratives (2002, 225). The first of these three relationships concerns the realistic or concrete portrayal of a community in which time is manifest in 'an organic fastening-down, a grafting of life and its events to a place, to a familiar territory with all its nooks and crannies, its familiar mountains, valleys, fields, rivers and forests, and one's own home' (225). This representation of time is reflected by a focus on communities and their experience of the world through a developing sense of local place and the cyclical changes that their environment undergoes. Alternatively, chronotopes may involve representations of the 'basic life-realities [that] are present in the idyll not in their naked realistic aspect [as an "organic fastening-down"] but in a softened and to a certain extent sublimated form.' These 'life-realities' include '[l]ove, birth, death, marriage, labor, food and drink, [and] stages of growth'; their appearance in 'sublimated form' may result in the representation of idealistic pastoral spaces poised in opposition to a world in conflict (226).

Sawyer explains Tom Shippey's contrast between the pastoral, understood as '"rural, nostalgic, [and] conservative,"' and 'fabril' literatures, which are '"overwhelmingly urban, disruptive, future-oriented, eager for novelty,"' and are centred on the image of the 'faber,' '"the smith or blacksmith in older usage, but now extended in science fiction to mean the creator of artefacts in general"' (2006, 402). Building on this collision between the pastoral and sf, Sawyer argues that '[m]uch of what I would call Pastoral in SF is the tension between these two modes: essentially a tension between enthusiasm for and anxiety about the future' (402). This mixture of anxiety and enthusiasm can be set against a contrast between fabril literatures and Bakhtin's craft-work idyll: both are in some sense concerned with fabrication and production, yet ideas of craft in the latter are perceived as wedded to a sense of tradition that is absent in the faber's preoccupation with creation and novelty. The influence of the pastoral on sf allows a contested space to develop where the productions of the faber are aligned with the craft-work idyll, thus recuperating ideas of technology from the potentially alienating perception of novelty. If the pastoral is understood as nostalgic, the faber or an analogue is already implicit in the tradition, as nostalgia assumes some novelty or change from which vantage the pastoral accrues its sentimentality. A similar dynamic is generated by aligning sf with the agricultural idyll, which brings ideas of terraforming into contact with the familiar domain of agriculture.

Apart from the active incorporation of overt themes into sf texts, Sawyer (2006) emphasises another intersection between the pastoral and sf that is centred on the internal literary effect generated by the use of specific pastoral techniques. These distinct engagements reflect different definitions of the pastoral and echo Leo Marx's (1964) sentimental and complex versions of the American pastoral. The sentimental pastoral ideal encompasses specific instances of landscaping and sublimation that generate pastoral nostalgia, which could be taken as a direct guide to political action. The pastoral design is a more complex ordering of meaning referring to 'the larger structure of thought and feeling of which the *ideal* is a part' (Marx, 1964, 24). The pastoral design adds complexity by introducing ambiguity into pastoral chronotopes; it manages 'to qualify, or call into question, or bring irony to bear against the illusion of peace and harmony in a green pasture' (25). It 'embraces some token of a larger, more complicated order of experience' by 'bring[ing] a world which is more "real" into juxtaposition with an idyllic vision' (25).

Sawyer connects these pastoral techniques to sf when he claims that 'SF/fantasy writers are reworking tropes once used by writers of the pastoral mode,' and he draws important analogies between the signifying practices of Elizabethan pastoral and the coding practices of sf and fantasy to argue that 'the pastoral, fantasy and science fiction are modes which reflect each other' (2006, 397). Drawing on William Empson's argument that the pastoral relies on a 'complex to simple' formula involving textual and linguistic strategies used to compress complex meaning into emblematic images (1950, 140), Sawyer connects the coding practices of the pastoral to the reading protocols of the sf megatext. Sf's iconic, 'emblematic' form of writing parallels the 'iconicity' of the pastoral emblem (2006, 404). The wilderness, garden and farm of the pastoral design are complex symbols, what Marx calls ecological images, and can be considered chronotopes insofar as '[e]ach is a kind of root metaphor, a poetic idea displaying the essence of a system of value' (1964, 42). In the terraforming narrative these ecological images intersect with sf discourse, offering conceptual handles for exploring the impact that technology has on human relationships to nature.

Sf narratives of terraforming, as stories of interplanetary migration, colonisation and adaptation, draw from aspects of the American pastoral. This tradition is rooted in images of America and of new lands depicted in literature, travellers' tales and colonial writing since the Age of Discovery (see Marx, 1964; and Kolodny, 1975). John Rieder has shown how sf shares these roots and contends that 'colonialism [...] is part of the genre's texture, a persistent, important component of its displaced references to history, its engagement in ideological production, and its construction of the possible and the imaginable' (2008, 15). The frequent references to and parallels between the terraforming of alien planets and the colonisation of America point to a cyclical unfolding of a specifically American experience. Colonial settlement of the American continent can be seen as an expression of utopian longing for a new beginning for communities aiming to create new socio-political foundations on which to flourish. The language of the American pastoral was a powerful motivating voice used to propagandise and raise support for colonisation. The utopian dimension of American colonisation was thus unstable: the new, unknown lands that promised better alternatives to the society at home could prove dystopian. The exploitation of indigenous inhabitants and colonial workforces meant that the construction of a pastoral utopia for the few was achieved at the expense of a dystopia for the many. Narrative treatments of terraforming draw from this convergence of the colonial and the utopian to inform their explorations of interplanetary settlement.

Representations of alien planets as wilderness landscapes often utilised what Ernest J. Yanarella has described as the Northern 'Garden of the Covenant' experience, which drew on Puritan traditions and parallels between the colonisation of America and the Exile of Israel from Egypt to motivate 'strenuous work and instrumental means to make way for a New Heaven and a New Earth' (2001, 80–81). The potential transformation of these wildernesses into agricultural land or Edenic gardens directed pastoral longing into the future, thus establishing one of the utopian dimensions that feed into the impulse to terraform. The Southern American 'Garden of the Chattel' experience offered a utopian vision that involved landscaping America as an Edenic paradise 'where nature's bounties need only to be cultivated and harvested' (81). The labour that enables habitation is sublimated, '[f]or underlying the pastoral landscape of the Southern colonies was the patriarchal plantation society with its system of slavery' (81). These two garden images eventually shifted towards representations of the Midwest 'Garden of the World,' which gathered into its orbit images of the pioneer farmer and an agrarian tradition that stimulated continued westward expansion. The 'imperialistic impulses of the pull of the Farther Landscape of the Far West' eventually led, in sf, to images of 'a technological garden or engineered millennium of the cosmos' (82) where '[t]he Further Landscape is now outer space, even deep space; and, once again, the garden image serves as an image of imperial expansion and conquest' (105–06). The utopian promise embedded in landscapes of the garden and farm played a significant role in structuring the colonial experience of America and has continued to inform the language of terraforming narratives.

The third and final distinctive feature of Bakhtin's idyllic chronotope is the 'conjoining of human life with the life of nature, the unity of their rhythm, the common language used to describe phenomena of nature and the events of human life' (2002, 226). Along with the motif of reciprocity between pastoral spaces and individuals, this 'common language' encapsulates one aspect of Ursula K. Heise's eco-cosmopolitanism: that the politics of place must account for the 'more-than-human-world' (2008). Empson argues that the symbolic use of the pastoral has often been to address human contradictions within communities (1950, 119), an idea reiterated by Tom Moylan in the context of utopian literature when he claims that '[t]he absent course of history is made sense of partly through the operations of the literary texts, for they are symbolic acts that provide imaginary resolutions to real social contradictions' (1986, 30). Negotiation between anxiety and enthusiasm for the future in pastoral sf highlights a temporal contradiction at the centre of the terraforming narrative's social enquiry. The traditional pastoral opposition between the country and the city is subject to change in the context of a technologically advanced industrial-capitalist society, what Bakhtin characterises as 'a great but abstract world, where people are out of contact with each other, egoistically sealed-off from each other, greedily practical; where labor is differentiated and mechanized, where objects are alienated from the labor that produced them' (2002, 234). This sense of alienation calls for developing a new awareness of humankind's place in a global complex of social relations within a scientifically understood universe:

> It is necessary to find a new relationship to nature, not to the little nature of one's own corner of the world but to the big nature of the great world, to all the phenomena of the solar system, to the wealth excavated from the earth's core, to a variety of geographical locations and continents. In place of the limited idyllic collective, a new collective must be established capable of embracing all humanity. (Bakhtin, 2002, 234)

This call for new conceptualisations of a collective sense of place that escapes the boundaries of its local and regional confines to encompass global scales is, in contemporary literature, answered by sf and its use of the terraforming motif to comment on society and to imagine alternatives. This version of the pastoral is not limited to the global; sf explores the new socio-political and ethical relations that advanced technology places us in with regard to Earth and the cosmos. Terraforming narratives mobilise pastoral themes and structures to create spaces to reflect on the possible relationships between individuals, communities, and nature, and to negotiate between anxiety and enthusiasm for the future.

The Garden of the World in Early 1950s Terraforming Stories

Sawyer points to *The Martian Chronicles* as an example of pastoral sf and notes that the opening vignette explores 'ideas of the "small town" (second paragraph) transformed by technology' (2006, 401). These vignettes operate like pastoral emblems: in 'Rocket Summer,' two ideas of the country town hinge on the metonymic climate change caused by a rocket, which brings the already transformed Midwest town into contact with the technological icons of a wider world. The unseasonable warmth works analogously to Marx's counterforce and signals a displacement of the chronotope of the town into a future where the space of the 'small town' is interpenetrated by technologies with far-reaching regional and global effects. The counterforce, Marx's paradigmatic example of which is the whistle of a locomotive heard from the pastoral idyll of the 'Sleepy Hollow,' dramatises the impact of two distinct worlds of experience, or in Marx's words 'structure of thought and feeling,' by representing technology's intrusion into the pastoral landscape (1964, 25–26). In sf, the counterforce takes on a new character as traditional motifs signifying the intrusion of technology into the pastoral landscape become the locus for nostalgia. Sf futurity reassigns familiar technology a new valuation as symbols of a bygone, pastoral era. Marx's example of the locomotive whistle as a counterforce becomes an essential icon for a historically superseded technological complex that is integrated into the pastoral landscape. The Midwestern town, once envisioned as the encroachment of civilisation into the wilderness, is embedded into the landscape as an essential part of *The Martian Chronicles*' pastoralism.

A spaceport appears in 'The Taxpayer' as a possible means of escape from Earth, which the eponymous taxpayer believes is a 'terrible world [...] there's going to be an atomic war!' (Bradbury, 1958, 48). The pastoral opposition between the country and the city is restructured as an opposition between Earth and Mars, and the flight from the city to the country becomes the desire for a literal flight from Earth, '[t]o get away from wars and censorship and statism and conscription and government control of this and that, of art and science!' (47). The pastoral chronotopes and themes that Bradbury utilises in the context of sf futurity tie into the text's background a conflict between anxiety and pastoral idealism. Mars's fundamental strangeness and its perceived contrast to the 'ordinary Monday morning on the ordinary planet Earth' offer an estranged setting to explore fears of unbridled technological and bureaucratic excess (48). Mars becomes a locus of desire, with the taxpayer drawing on pastoral images of the planet as 'a land of milk and honey' (47). The foundation for the counterforce to this pastoral ideal has already been established by earlier stories that relate the failure of the first three scouting expeditions to Mars; spaceport security mentions the uncertain outcome of these missions in an attempt to deter the taxpayer from his desperate dream of a flight from Earth.

'The Green Morning' and 'The Locusts' show how sf works analogously to the pastoral design to reflect critically on terraforming as a mirror to American colonialism. In 'The Green Morning,' Benjamin Driscoll is inspired by the frontier legend Johnny Appleseed to plant Earthbound vegetation on Mars to transform the planet, not by 'making just fruit for the stomach, [but by] making air for the lungs' (Bradbury, 1958, 97). He sees this act of terraforming in militaristic terms as a 'fight against the very thing that might prevent his staying here. He would have a private horticultural war with Mars': by aligning gardening with war, one of the central problems motivating the settlers' journey, Driscoll landscapes Mars in terms of the Garden of the Covenant ideal (98). This story's counterforce stems, not from Earth's civilisation, but from Mars's alien otherness: '[t]here lay the old soil, and the plants of it so ancient they had worn themselves out. But what if new forms were introduced? Earth trees [...] There was no guessing what mineral wealth hid in the soil, untapped because the [...] [Martian] trees has tired themselves to death' (Bradbury, 1958, 98). This reference to mineral wealth can be understood as a comment on the chemical value of certain minerals to Earth's vegetation, but it also compacts within the trope of dying alien life the theme of the colonial acquisition of natural resources and histories of the American gold rush. This signification recasts the pastoral ideal that Mars offers into a form of colonial invasion. Landscaping Mars as an American frontier foregrounds the conflict between the past and future embodied in the opposition between a nature on Mars that derives its value from its age and a culture perceived in opposition as exuberant and progressive and therefore a suitable heir to its wealth.

The Martian Chronicles portrays terraforming as the usurpation of Martian life in a complex of associations where Mars functions as a sign of futurity as well as anachronism, understood as 'an incongruous co-habitation of the same moment by people and artifacts from different times' (Rieder, 2008, 5). Mars is variously positioned as a locus of primitivist pastoral desire and a mirror to civilisation. 'Mars was a place as unpredictable as time'; the occurrence of an impossible event within the established sf frame, the overnight growth to maturity of the seeds that Driscoll plants, is an emblem of the pastoral promise of the Garden of the World (Bradbury, 1958, 99). This compression of time allows the Mars chronotope to operate as an iconic site of projection that hosts a confrontation between the future and past. Driscoll uses the emblem of the fertile Martian soil to figure the nurture of American pastoral ideals, notions that are entangled with colonialism. He 'thought of the rich, inky [Martian] soil, a soil so black and shiny it almost crawled and stirred in your fist, a rank soil from which might sprout gigantic beanstalks from which, with bone-shaking concussion, might drop screaming giants' (Bradbury, 1958, 99). Deeply ironic, the 'giants' of this metaphor figure the colonists themselves, who 'fall' from Mars when they decide to return to Earth after the long-predicted outbreak of nuclear war described in the stories 'The Luggage Store,' 'The Watchers' and 'The Silent Towns.' At a further level of abstraction the metaphor stands as an emblem of hubris: humanity's faith in technology is undermined because it chooses to cultivate misplaced ideals. The Martian landscape figures the raw potential for the realisation of the desires that the colonists choose to nurture in this new space, but it also possesses a dual aspect that threatens a reversal of the pastoral utopianism that the colonists attempt to grasp: the soil is rank and, like the carapace of an insect, black and shiny; thus the ideal of the rich and fertile landscape anticipated by the colonists is overturned. The 'inky' soil suggests that Mars is a space that can be overwritten by the colonists, a palimpsest where a new history can be devised. However, the dynamism and seeming agency of the landscape, which exhibits almost lifelike qualities ('crawled,' 'stirred'), works to rewrite the role of the colonists in this 'Jack and the Beanstalk' image of terraforming.

Until the impossible occurrence in this story, this image only leans towards the literal. Its emblematic effect is modulated by combining the chronotope of the Martian landscape with a temporal dimension expressive of utopian desire. The counterforce's most prominent work is to show how, despite Driscoll's attempt to adapt the planet to the settlers, the environment itself works a physical change that gestures towards an alternative to his landscaping of Mars. Driscoll faints when first arriving on the planet, a sign of his difficult adaptation to its alien otherness, and suffers a repetition at the climax of the story when he takes 'one long deep drink of green water air'; he 'felt his rib case. In thirty days, how it had grown. To take in more air, they would all have to build their lungs. Or plant more trees' (Bradbury, 1958, 100). The counterforce introduces a choice between modifying the landscape and adapting to the new environment, both physically and ideationally.

'The Locusts' counters the potential for internal change explored in 'The Green Morning.' This vignette begins with the arrival of rockets which 'made sand and silica into green glass which lay like shattered mirrors reflecting the invasion, all about' (Bradbury, 1958, 101). These green mirrors symbolise Earth's eclipsing of Mars's otherness. Drawing from the Southern Garden of the Chattel theme of the pastoral landscape as fertile for both life and the imagination, this eclipsing results in the cultivation of attitudes and desires that ironically threaten a re-enactment of the circumstances motivating the settlers' flight from Earth. Mars's potential as a pastoral retreat depends on its otherness, the very aspect that is diminished by further colonial waves: 'from the rockets ran men with hammers in their hands to beat the strange world into a shape that was familiar to the eye, to bludgeon away all the strangeness' (101). Mars's alien otherness is also entangled with notions of racial and cultural difference. In 'The Shore,' we learn that '[t]he second men should have traveled from other countries with other accents and other ideas. But the rockets were American and the men were American and it stayed that way [...] the rest of the world was buried in war or thoughts of war' (111). In 'Way in the Middle of the Air,' a black exodus to Mars, opposed by the white racist Samuel Teece, punctures notions of an idealised pastoral town by recalling histories of slavery and lynchings underwriting the Garden of the Chattel images of the Southern American colonial experience (Yanarella, 2001, 81). Bradbury's pastoral sf warns of the dangers of landscaping Mars according to an ideal that Marx argues 'was embodied in various utopian schemes for making America the site of a new beginning for Western society' (1964, 3). The utopian landscape that terraforming seems to promise is overshadowed by an ambiguous oscillation between the chronotope of an alien Mars and Mars as a duplicate America that reflects contemporary social experience.

'The Million Year Picnic' maps time to construct a primitivist distancing from Earth and the values that it embodies. As war leads to Earth's final destruction, a family escapes '[a] million years' into the now deserted Mars (Bradbury, 1958, 213). They travel back in time, away from civilisation's legal, economic and political structures, but above all its technocracy:

> I'm burning a way of life, just like that way of life is being burned clean of Earth right now. [...] Life on Earth never settled down to doing anything very good. Science ran too far ahead of us too quickly, and the people got lost in a mechanical wilderness, like children making over pretty things [...] emphasizing the wrong items, emphasizing machines instead of how to run the machines. (Bradbury, 1958, 220)

The creation of 'mechanical wildernesses' can be usefully read against Sharona Ben-Tov's argument that '[w]e invest in technology as a means of replacing fallen nature, fixing the past, and returning to the American Earthly Paradise' (1995, 55). Terraforming is thus a method for creating via technological means anachronistic worlds rooted in the pastoral ideal. Despite Ben-Tov's claim that because 'science fiction's commitment to the ideologies that produced modern science and technology [is] built into its very structure, there are idols that it cannot break, ideas and attitudes that it cannot subvert,' the terraforming motif, as exemplified by Bradbury's pastoral sf, makes use of the megatext and elements of the pastoral mode to generate degrees of ambiguity, ironic detachment or reflection towards these ideas and attitudes (51). Anxiety and enthusiasm for the future are established through strategies that mirror techniques central to the pastoral design. *The Martian Chronicles* comments on the changes that technology enacts on society while highlighting a complex of concerns experienced in the international context of postwar America, issues that include 'politics, the atom bomb, war, pressure groups, prejudice, [and] laws' (Bradbury, 1958, 165).

Bradbury also wrote a proto-Gaian living world story published separately from *The Martian Chronicles* that deploys sentimentalised pastoral elements within a larger design to reflect on the ways technological societies disrespect nature. 'Here There Be Tygers' (1972 [1951]) refigures the theme of identification with a living world and offers a transformative revelation of nature's cosmological otherness. Like Murray Leinster's 'The Lonely Planet' (1949), Bradbury's story presents a world where the desire for utopia can be satisfied only if Promethean fears of nature's otherness can be overcome. A group of interplanetary explorers chartered by a resource extraction company encounter a world that grants them their unconscious wishes. The crew see a pastoral utopia whose grass 'was the freshest green colour they had seen since childhood' (Bradbury, 1972, 120). The colour green is central to Bradbury's utopian discourse, and its use here is an invitation to the colonists to reflect on their lost childhood: '[r]emember how you used to run when you were a kid, and how the wind felt. Like feathers on your arms. You ran and thought any minute you'd fly, but you never quite did' (122). Unexpectedly (for the colonists), Driscoll does fly, the first indisputable evidence that the planet has been directing itself towards their desires. The landscape is ideally suited to leisure: '[t]he men laughed quietly in the baseball season, in the good quiet wind for tennis, in the weather for bicycling and picking wild grapes' (122). It is nostalgic and utopian, prefiguring Kim Stanley Robinson's treatment of softball in *Pacific Edge* (1990), but it is also homocentric, highlighting a masculinist orientation to the pastoral utopia. Nature's otherness is effaced as the planet mirrors the expedition's fantasised notions of childhood.

Chatterton, the company's representative, sees a devious world whose games are not the innocent leisure activities of youth. For him '[i]t's too green, too peaceful,' and his anxiety towards the planet is apparent when he asks '[h]aven't you *felt* it? This world's alive, it has a look to it, it's playing with us, biding its time' (Bradbury, 1972, 121, 123–24). He projects an exploitative orientation to the planet and brings guns and an 'Earth Drill' to support their conquest with violence, repeating the history enacted on Earth:

> 'You have to beat a planet at its own game,' said Chatterton. 'Get in and rip it up, kill its snakes, poison its animals, dam its rivers,

> sow its fields, depollinate its air, mine it, nail it down, hack away at it [...] You can't trust planets. They're bound to be different, bound to be bad, bound to be out to get you.'
>
> [...]
>
> Earth was far away, her system and her sun forgotten, her system settled and investigated and profited on, and other systems rummaged through and milked and tidied up. (Bradbury, 1972, 119)

These tropes relate the image of the drill and mining to masculinity and symbolic sexual penetration. Chatterton, however, contradicts the impression that '[i]f ever a planet was a woman, this one is' by masculinising it: '[w]oman on the outside, man on the inside [...] [a]ll hard underneath, all male iron, copper, uranium, black sod. Don't let the cosmetics fool you' (Bradbury, 1972, 120). He attempts to counter the symbolism of Earth as woman or mother by identifying masculinity with the planet, thus justifying his antagonism. Conversely, the others respond favourably to the landscape, 'like very young men in the presence of great beauty, of a fine and famous woman' (125).

Koestler speculates that the world is like '"[a] woman who'll do anything to please her guests, as long as we're kind to her"' (Bradbury, 1972, 129). Resonating with Leinster's portrayal of the living world Alyx, the feminised otherness of the landscape in 'Here There Be Tygers' caters to the wish fulfilment fantasies of the male expedition. One of the text's turning points is a dream in which the crew are informed of a female population inhabiting the planet, a strong motivation for the expedition to stay. Gendered identity becomes a key ideological space in which the text's environmental discourse attempts to negotiate relationships to the landscape in terms of human sexual politics. The theme of resource exploitation recapitulates Arthur Conan Doyle's gendered use of the drill in 'When the World Screamed' and is presented as a moral transgression, with the narrator explaining that '[s]he [the planet] wanted to be loved, like every woman, for herself, not for her wealth' (132).

Robert Heinlein's *Farmer in the Sky* and Arthur C. Clarke's *The Sands of Mars* are the first novels to focus their narratives on terraforming. Responding to the same concerns expressed in *The Martian Chronicles*, they utilise chronotopes that appear in many later terraforming narratives, positioning them in relationships that mirror the pastoral country/city opposition to build worlds that engage with the issues involved in constructing new social relationships. They make use of specific coding practices to compress meaning into various chronotopes and of specific allusions to the colonisation of the American continent, through both the narrative representation of worlds and the characters' orientation to them. While it is certainly true that images of the frontier have dominated terraforming stories, they do not simply recapitulate the various aspects of the frontier (although they do this too) but refigure and reorganise them in accordance with the narrative and linguistic possibilities offered by sf discourse.

Farmer in the Sky is an early juvenile that embodies a semi-primitivist orientation towards the past. The colony's division into 'homesteaders and townies,' the latter made up of the planet's professional, political and administrative members, clearly mirrors the pastoral country/city opposition (Heinlein, 1967, 83). Archetypal frontier heroes Johnny Appleseed and Daniel Boone are frequently alluded to (58), along with the American tradition of Scouts established in the spirit of exploration that Boone came to symbolise (36). This allusive strategy is also present in a more sophisticated form in Bradbury's portrayal of a phantasmagoric Mars and appears with regularity throughout the terraforming tradition. Also significant are comparisons between the colonial ship the *Mayflower* and its futuristic double (the colonial spaceship; 77), scattered references to Christopher Columbus (22), and the Ganymedean settlement's direct parallel to a frontier town (60). Other pastoral motifs include the song 'The Green Hills of Earth,' which features lyrics praising the rocket in terms suggestive of the pastoral hero's journey and return, with new insights, to the city: *'Out ride the sons of Terra; Far drives the thundering jet –* [...] *We pray for one last landing on the globe that gave us birth –'* (55). These direct allusions belie the text's sf futurity and evidence the colonists' landscaping of Ganymede as a new frontier in the tradition of the American pastoral. This landscape is overlaid with others in the sf world that Heinlein builds, which incorporates the pastoral complex of value into a new socio-political context reflecting contemporaneous postwar anxieties.

Paul du Maurier, the leader of a scouting party prospecting for new areas to settle, recalls Bradbury's use of global warfare in *The Martian Chronicles* but draws from 'mathematical population bionomics' (ecology) to predict that a Malthusian population increase, resulting in atomic war on Earth, is between forty and seventy years away (Heinlein, 1967, 154–56). Heinlein circumscribes the pastoral ideal within a framework that connects the scientific principles underpinning ecology to farming and statistics in ways central to the terraforming tradition. These themes demonstrate the continuing influence of energy economics on terraforming narratives, a dynamic that can be traced back to Wells's presentation of a global society managed by a scientific elite. Paul's prediction of war, part of the 1950s consensus future constructed by many works of sf at the time, operates in an ideological space alongside ideas of migration and autarky. One scout asserts that 'Ganymede has got to be made self-sufficient as soon as possible – and then we've got to slam the door!' (153). Paul argues that while self-sufficiency is important, immigration will not disturb political and economic stability on Ganymede simply because Earth will lose the capability for space travel in the certain event of global warfare: 'Building colonies. We [the Earth Commission] think that is worthwhile in itself. The colonies need not be affected by the War. In fact, I don't think they will be, not much. It will be like America was up to the end of the nineteenth century; European troubles passed her by' (Heinlein, 1967, 155). Characteristically, interplanetary war is understood in terms of American colonial history by virtue of an apparent geographic analogy. Anna Bramwell argues that the influence of peasant ideology in Europe and America on the energy economists of the early twentieth century was a significant factor in shaping views of the land, which came to be seen as a fixed resource that would be most efficiently worked by peasants on small farms, rather than by larger farms overseen by landowners. Furthermore, Bramwell notes that '[p]easants were seen as the source not only of social cohesion and conservative values but of ecologically sound agricultural improvement' (1990, 66–67). Ganymede, with its promise of free land, offers not only a landscape of untapped resources but an opportunity for the construction of an economy and a political system based on the myth of the traditionalist peasant farmer, thus bypassing the geopolitical tensions that have developed on Earth. This rewriting of history taps into the Garden of the Covenant ideal to construct an image of the self-sufficient and independent colony sealed off from its centre by an unbridgeable vacuum of space. Portrayals of terraforming in *Farmer in the Sky* foreground desires for the creation of autonomous worlds where history can diverge from the imagined future envisaged in the text.

The libertarianism of *Farmer in the Sky* gains its impetus from ideas of entropic processes and their effects on energy economics, which highlighted the finiteness of Earth's resources. The political implications of self-sufficiency – isolationism – are justified by recourse to an ecological awareness of the material limits that Ganymede imposes on colonisation. The motivating impulse to terraform Ganymede is to escape the resource-depleted and overpopulated Earth, for which food rationing operates as a synecdoche. Bill Lermer reflects that, '[l]ike most everybody, we had come out there on the promise of free land and a chance to raise our own food' (Heinlein, 1967, 83). He complains about the global management of food on Earth in a manner that resonates with some attitudes towards modern anxieties about climate change: '[w]hat's the use in being careful if somebody on the other side of the globe is going to spoil your try? Those darned Chinese ought to quit raising babies and start raising food!' (Heinlein, 1967, 9). Reflecting fears over the increasing internationalisation opened up as a consequence of postwar capitalist expansion, this episode highlights the limits of a global distribution of finite resources and establishes the background for the narrative's advocacy of a pastoral localism made possible through terraforming. Later, Bill describes a colonial farm on Ganymede that, unlike productive America, 'remains as primitive as a Chinese farm' (Heinlein, 1967, 107). On Ganymede, such a relationship emphasises the human labour involved in returning to an agrarian economy based on the image of the colonial farmer. On Earth, China's burgeoning population operates as an icon for Malthusian overpopulation; its lack of a technological (agricultural) infrastructure establishes it as an anachronism, pointing towards an extrapolation from essentialist ideas of primitive cultural otherness.

In contrast to rationing and the artificial 'Syntho-Steaks' of Earth, Bill's narrative revels in the abundance of food on Ganymede, which includes '*real* ham' and reminds him of Scout camp (Heinlein, 1967, 8, 78–79). Ganymede is no land of Cockaigne, however, but a wilderness that draws more from the pastoral Garden of the Covenant ideal. Marx's comments on the utopian garden image's 'ancillary notion of the new continent as a land of plenty' may explain the dynamics of this contrast, for he argues that the sixteenth-century voyager's notion of '"incredible abundance"' is 'perhaps the most important single distinguishing characteristic of American life. In our time, to be sure, the idea is less closely associated with the landscape than with science and technology' (1964, 40). Anxiety over the capacity of science and technology to supply the abundance that in the past appeared as a fundamental facet of American culture is represented as Earth's failure, in an international context, to supply this abundance to a global population. While the technology that enables terraforming offers a technical solution to this problem, the ideological connotations of a new colony that harks back to the pastoral ideal of an American frontier are coherent with a desire for a return to the superior productive efficiency of the small farming community.

The various trials through which Bill and the colonials must prove their worth – as cultural assets, as in the episode where Bill demonstrates the cultural worth of his accordion to a committee, as a Scout on Ganymede, as a settler looking to 'prove' land by converting the rocky wilderness into fertile agricultural space, and as a member of a community who, in the face of disaster, remains to rebuild – function as tests of character and evolutionary potential. These trials foreground the role individuals play in creating an independent community worthy of survival. Individuals who unrealistically depend on government promises to support them, or who attempt to manoeuvre Earth into making impossible provisions for the colony, demonstrate their inability to adapt to the new socio-political demands on Ganymede and leave. When he considers returning to Earth with other colonists after a quake kills a third of their community, Bill has in mind those who choose to leave when he tells his father, 'I don't *like* being classed with these lugs' (Heinlein, 1967, 22). This episode, in which undesirables are winnowed from the colony, valorises the blend of rugged individualism and community spirit that Bill demonstrates by choosing to stay. His struggle towards self-supporting agrarianism mirrors the colony's struggle for self-sufficiency, while the speed with which he reconciles himself to the difficulties of habitation on Ganymede affirms the significance of the colony's own existence. If Paul's claim that atomic war is inevitable is indeed right, the colonists who leave simply damn themselves to extinction. According to the logic of Social Darwinism established in the text, the values held by Earth's civilisation are evolutionarily unviable while the colony's brand of American colonialism is purportedly otherwise.

Ideas of Social Darwinism are implicit as an explanation for the difficult adaptation of the organism to an unfamiliar environment. Bill's young stepsister Peggy is unable to acclimatise to Ganymede, a development that threatens to make necessary a return to Earth for the family. Peggy, however, dies shortly after being injured in the quake, thus underscoring notions of fitness to habitation involved in terraforming. Bill uses an analogy that relates plants to people, asking, 'Have you ever had a plant that refused to be happy where you planted it? It was like that. She belonged back on Earth' (Heinlein, 1967, 123). This megatextual compression of meaning demonstrates a convergence between the pastoral 'complex into simple' formula, pastoral themes (the plant analogy itself) and sf oppositions between planets that stand for political positions and ideas that are structurally analogous to the pastoral country/city dynamic. This image parallels the gardener's cultivation of space and the planning of a new society based around the family unit. Significant here is the level of agency ascribed to Peggy in 'refusing' to be happy, a case in which the gardener's plans are thwarted by conditions outside of his control. The corollary of this analysis is that the gardener cannot be censured for a failure of the individual plant to thrive. Peggy is uprooted from her environment by the demands of a larger organisational unit, but it is not the fitness of the environment to the individual that is at fault but the individual's failure to adapt to the social and physical realities of Ganymede.

When Peggy dies, Bill reflects that if he had not encouraged her to stay, she would be 'right back in California, not here in this damned place where she couldn't live, where human beings were never meant to live' (Heinlein, 1967, 145). Located at the natural boundary of the Pacific Ocean and thus preventing egress for the continued westerly expansion of the pioneers who had travelled across the American continent, the Californian setting is significant as the origin of the new pioneers to Ganymede. That California is metropolitan rather than rural or wilderness underscores the centrality of the urban space as a secure home in contrast to the dangers of Ganymede. Bill's pessimism in the light of Peggy's death, his assertion that humankind was 'never meant to live' on other planets, suggests that in order for the colonists to successfully make a home of Ganymede, they must struggle with and transcend the limits of their humanity. Peggy's death suggests that this failure to adapt to other planets indicates a more fundamental inability to adapt to a future where interplanetary colonisation is the only answer to war and extinction on Earth. Those colonists who were 'meant to live' are, in Social Darwinist terms, those whose individual qualities are suitable adaptations to the social and physical requirements of the Ganymedean settlement. The contours of an ideology of manifest destiny connecting expansion throughout space to the transcendence of human limitations are thus sketched.

The text's appropriation of elements of the Garden of the Covenant ideal suggests that survival in spite of the hostility of Ganymede's environment is a triumph of individual and social endeavour, and it elevates the type of individual and community who successfully inhabit Ganymede to a pastoral, heroic status. The colonials' farms are less technologically sophisticated than the material infrastructure of Earth, and yet technology is central to terraforming Ganymede (bulldozers and spaceships play an essential role). This semi-primitivist distancing from the technologism of Earth offers a space to dramatise the need for the appropriate distribution of resources amongst the community, which in turn is imagined as a method for forging stronger bonds between individuals based on the necessity of co-operation for survival.

In a passage that recalls Driscoll's dilemma in Bradbury's 'The Green Morning,' one character asks:

> What are we going to make of this planet? We can make it anything we want. Mars and Venus – they had native cultures. We dare not

> change them much and we'll never populate them very heavily. These Jovian moons are another matter; it's up to us. They say man is endlessly adaptable. I say on the contrary that man doesn't adapt himself as much as he adapts his environment. Certainly we are doing so here. But how? (Heinlein, 1967, 152)

The emphasis in *Farmer in the Sky* is not on the material means of terraforming, but on how the use of technology within a community affects individuals and how these experiences scale up to the level of the community. If humankind is infinitely adaptable, notions of a stable human identity are subverted, and yet the community portrayed undeniably reflects concepts of American national identity prevalent in the 1950s. The pastoral ideal offered here suggests that smaller, local communities and libertarianism, in contrast to a rigidified bureaucratic system of global resource management on Earth, offer viable answers to international economic structures that manifest inequality in the very attempt to cater fairly to the global population. The question of adaptation raised by Bradbury – which calls for a fundamental change to human nature – is supplanted in *Farmer in the Sky* by the question of how the environment itself is being transformed. The text, however, uses this focus on the agricultural value of space to suggest that a pastoral return to a simpler, American past offers an antidote to the problems and anxieties raised by the extension of democracy to the world.

Echoing Bill's journey to Ganymede in *Farmer in the Sky,* Clarke's *The Sands of Mars* (1976) opens with the journey into space of sf author Martin Gibson, whose stories have been superseded by the reality of Mars's colonisation. Gibson's surprise when first meeting the professional astronauts piloting the spaceship *Ares,* whose ordinariness 'was not fitting at all well into the expected pattern,' highlights the constructedness of Earth's landscapes of space and the Martian colony (17). Explicit comparison of the astronauts to American pioneers is belied by their actuality: '[t]here was no way of guessing that they belonged to a profession more romantic than any that the world had known since the last cowboys traded in their broncos for helicopters' (16). The re-use of tropes from sea narratives in representations of air travel in stories such as Rudyard Kipling's 'With the Night Mail' (2009 [1905]) also had an analogue in stories of space travel, which were often based on 'the assumption that there would be no fundamental difference between the ships of space and ships of the sea – or between the men who manned them' (Clarke, 1976, 30). Space travel is thus envisaged as sea travel, implying that the colonisation of America parallels that of alien planets. *The Sands of Mars* takes this traditional sf basis for imagining space colonisation and attempts to explore the dissonances raised by the American pastoral in sf. It engages in an explicit dialogue with the conventions of the 1950s consensus futures in an attempt to negotiate some of these distinctions. By focalising events through the consciousness of an sf writer, Clarke develops a meta-commentary on the science-fictionalisation of (science-fictional) events and self-consciously appraises the refiguration of the pastoral in ways absent in *Farmer in the Sky*.

The Sands of Mars explores the important differences between sf and pioneer traditions, noting the discrepancy between romanticised expectation and the reality of astronautics and the Mars colony. During an interview with Mars's Mayor Whittaker, Gibson questions the validity of understanding interplanetary colonisation in terms of historical analogues. Whittaker's response is that '[i]t can't be pressed too far. After all, men could breathe the air and find food to eat when they got to America!' (Clarke, 1976, 90). *Farmer in the Sky* also recognises these fundamental material differences in the accessibility of resources, despite the amelioration of these basic facts by the narrative's modelling against themes related to American colonisation (Heinlein, 1967, 85). The relocation of such elements onto other planets highlights a contradiction between the past and the future implicit in the American pastoral's reworking within the textual space of the terraforming narrative. This contradiction reflects the problem of using past colonial narratives as guides to understanding and responding to the planet; the tension between these landscapes and the colonists' lived experience of Mars points to a temporal circularity that underwrites their vision of the future.

The image of the garden is used to playfully undercut the symbolic import of a ceremony held to commemorate the cultivation of an indigenous Martian '"airweed."' The inauguration of this new method for increasing the planet's atmospheric oxygen is transmitted to Earth, and Mayor Whittaker and Chief Executive Warren Hadfield take the opportunity to propagandise for the terraforming project. Gibson thinks that the airweed 'certainly didn't look as if it could control the future of a planet' (Clarke, 1976, 192), while the narrator (focalised through Gibson's perspective) remarks that 'Hadfield had finished his token gardening: someone else could complete the job and fill in the hole. (The planting team was already hovering in the background, waiting for the big-wigs to clear out of the way so that they could get on with their work.)' (Clarke, 1976, 192). This narrative technique plays on Gibson's status as an outsider who, in his official journalistic role, views the ceremony with a detached and ironic gaze, thus raising a discrepancy that the text attempts to bridge between the engineering essential to portrayals of terraforming and the process of gardening. These landscapes are oriented towards the future and to the transformative potential symbolised by the Martian plant, itself able to survive only with difficulty in this unforgiving environment. Like the impossible event in Bradbury's 'The Green Morning' and the agricultural cultivation of land in the 'Johnny Appleseed' chapter in *Farmer in the Sky,* this event operates as an emblem that activates an allusive network of pastoral associations within the futurity of the imagined world.

The opposition in Clarke's novel between Earth and Mars is more optimistic in tone than in Bradbury or Heinlein's stories, yet relations between the newly discovered Martians and the colonists retain a distinctly imperial aspect. Unlike Bradbury's and Heinlein's works, which portray the economic stresses caused by conflicts internal to Earth, *The Sands of Mars* (1976) focuses debate on Earth's continued economic support for the terraforming project rather than on portents of global warfare. Nevertheless, a political and technical struggle for self-sufficiency structures the plot, providing the motivation behind 'Project Dawn' (the technical transformation of Phobos into a sun) and Gibson's own contribution to the colony – his attempt to raise awareness on Earth of 'the spirit we've built up here on Mars' (169). As in *Farmer in the Sky,* notions of community establish one axis for the contrast between Earth and Mars:

> Here were men and women united in a single task, driving towards a common goal, each knowing that their work was vital to the community. They had a sense of fulfilment which very few could know on Earth, where all the frontiers had long ago been reached. It was a sense heightened and made more personal by the fact that Port Lowell was still so small that everyone knew everyone else. (Clarke, 1976, 118–19)

Unity of motivation and action are central to this sense of community spirit, which is driven by an internalised ethic of expansion that motivates the decision to terraform. The continued expansion of the frontier hinders the development of a local, personal sense of connection to the community. Terraforming appears to offer a solution to this problem, a pastoral flight from Earth and a return to small communities able to exercise political and social self-determination.

Like Heinlein's text, *The Sands of Mars* employs aspects of the family idyll and combines them with a romantic subplot (the love idyll) to underline Gibson's growing investment in the future of the colony.

Gibson is integrated into the life of the local community as a result of a change of consciousness. A promised union between his recently discovered son and the daughter of the colony's chief executive stands as an emblem for this new orientation towards the future. If Gibson's official acceptance into the colony and his growing bond with his son leave him with the feeling that '[f]or the first time in his life, [he] had a future to which he could look forward with interest and excitement – a future which would not be merely a repetition of the past,' the text's commentary on sf's colonial inheritance leaves such repetition in tension with the movement towards new trajectories for humankind's future (Clarke, 1976, 196).

A significant intersection between colonialism and the pastoral is developed through the colony's discovery of sentient Martian life. Early evidence of the capacity for these 'beasts' to learn to perform basic tasks, while not proof of intelligence, suggests the potential for their evolutionary development. A new motive joins the terraforming project, that of adapting the planet to create ideal conditions not only for the colony, but for the Martians to thrive: '[t]here was something inspiring in the thought of regenerating not only a world, but also a race which might be older than man' (Clarke, 1976, 200). Echoing the dying Martian life in Bradbury's 'The Green Morning,' the regeneration of life through terraforming attempts to justify the habitation of Mars and implies that the colony is able to determine the best interests of the Martians and their environment.

Human–alien relationships are restricted to interaction with Squeak, '[o]ne baby, less than a metre high, [who] could only be described by the overworked adjective "cute"' (Clarke, 1976, 150). Unlike the adults of its species, it appears unusually curious and follows Gibson back to the colony. Although he 'sometimes felt rather like a baby-snatcher who had abandoned his victim immediately after stealing it,' Gibson feels no qualms over making use of the aliens to cultivate Mars (172). Patrick D. Murphy notes that 'these animals and plants are portrayed as oddities rather than as living components of a biosphere' and that 'the colonizers only care about adapting it [the planet] to terraforming' (2009, 376). Instead of a pantropic adaptation to the ecology of Mars, terraforming is seen as the preferred method of settling alien planets.

After establishing the chronotope of the colony as a regenerative space where time unfolds as a process of personal healing and socio-political reconstruction and self-determination, the narrative turns its attention to resolving anxieties over colonial relationships with indigenous cultures by invoking the theme of the discovery of alien civilisations:

> For it was their world, not Man's. However he might shape it for his own purposes, it would be his duty always to safeguard the interests of its rightful owners. No one could tell what part they might have to play in the history of the universe. And when, as was one day inevitable, Man himself came to the notice of yet higher races, he might well be judged by his behaviour here on Mars. (Clarke, 1976, 200)

The idea of contact with 'higher' alien civilisations and the external evaluative role that they might play function as a check to the colony's ethical standards. Hadfield reminds the colonists that they are 'making history,' creating the foundations for a future of terraformation to which other generations of humankind will be indebted (Clarke, 1976, 188). The first experiments with terraforming therefore involve attempts to formulate sound ethical relationships within and between species. These ethical considerations underpin a colonial project extending beyond the solar system that struggles to escape a repetition of the record of oppressions observable in Earth's own history.

The colony's recourse to terraforming as an answer to Earth's reluctance to provide economic support for colonisation is hindered, in Gibson's view, by Earth's own anxiety over the potential loss of its status as imperial centre of a new interplanetary society. Gibson speculates that '[i]t's too wounding to their pride. They want the Earth to remain the centre of the universe,' and posits the colony's independence as a threat to Earth's self-image. Norden, one of the astronauts shuttling between Earth and Mars, pinpoints the homogenisation at work in Gibson's theory, noting how 'it's funny how you talk about "Earth" as if it were some combination of miser and bully, preventing all progress here,' while admonishing Gibson not to 'forget that everything you've got here is due to the enterprise and initiative of Earth' (Clarke, 1976, 202). Norden's position as an intermediary between the two planets gives him an insight into the root of the conflict between Earth and Mars and informs his observation that the colonists 'take a very self-centred view of things' (202). Norden's statement shows how Mars's relationship to Earth is mediated by the interests of those directly administering policies for the colonial government.

One factor preventing a synthesis between the two planets is '[t]he sheer difficulty and expense of interplanetary travel,' a condition that inevitably leads to 'some lack of understanding, even intolerance, between Earth and Mars.' These physical conditions encourage a corresponding mental orientation, prompting Gibson to hope that 'these psychological barriers would be broken down and the two planets would come closer together in spirit as well as in time' (Clarke, 1976, 202). The dynamics of this reconciliation mirror to some extent the American pastoral's garden chronotope, which synthesises the emblematic value of the city and the wilderness. Here, the technology of space travel effects a bridging of opposites. The Mars colony, represented as a garden to be extended across the Martian wilderness, is an attempt to forge a new culture where values specific to Mars can be developed independently before being fed back to Earth. The political reconciliation that Gibson imagines highlights social barriers that emerge as a consequence of the realities of physical space and colonial landscapes, barriers that must be overcome through increased contact between Earth and an autonomous Martian colony able to exercise self-determination.

Bradbury's, Heinlein's and Clarke's terraforming narratives foreground pastoral chronotopes and themes of the 'sentimental' variety but imbricate them with megatextual sf tropes. In the process they position terraforming in an unstable zone where oscillation between the past and future speaks of an anxiety towards an imagined technological adaptation of Earth that is driven by political and economic changes in contemporary society. The dissonance between each text's treatment of themes drawn from the American pastoral and the alien status of the planets undergoing terraforming creates a sense of estrangement from Earth's socio-political norms. The similarities between the imagined Earths in these works are evidence of a consensual future constructed dialogically by the sf of the period, a future that later writers such as Pamela Sargent and Kim Stanley Robinson write against in their constructions of Earth and interplanetary colonies. Bradbury, Heinlein and Clarke use terraforming to express different orientations towards the colonial expansionism that informs globalisation. While Bradbury and Heinlein imagine terraforming as a reaction to fears that dwindling resources and an increase in geopolitical contact may lead to global warfare on Earth, Clarke's use of terraforming acknowledges the barriers to reconciliation that prejudice erects but expresses optimism that the future offers an escape from a repeating human history.

The Burden of Hope in the Garden of the Chattel: 1950s Consensus Dystopias

Frederik Pohl and C.M. Kornbluth's *The Space Merchants,* Walter M. Miller's 'Crucifixus Etiam' and Poul Anderson's 'The Big Rain' and *The Snows of Ganymede* construct models of dystopian societies that are contrasted with images of Earth to highlight the political and social exploitation underlying the formation of new societies. In this regard they resemble Bradbury's depiction of terraforming in *The Martian Chronicles*. Whereas Clarke's and Heinlein's works depict a sense of collective movement towards the instantiation of pastoral landscapes, these dystopian texts locate the pastoral golden time in a sealed-away future inaccessible to the colonists. Social arrangements on the colonised planet are frozen in a nightmare cycle of entrapment; these stories expose the exploitation of the inhabitants' labour by a colonial elite, recalling the Garden of the Chattel images of the Southern American pastoral experience of colonisation. Although the presence of romantic pastoral motifs of the pioneer tradition is downplayed, such relations are not entirely absent from this group of texts. Examples of sentimentalised pastoral themes are incorporated into the political design of each in a way that works analogously to Marx's pastoral design, allowing them to reflect and comment on social exploitation. Earth is established as a political contrast, with sentimental themes appearing only to underscore the inequities that underlie vast societal projects. The imaginary construction of sf landscapes via textual world-building allows additional symbolic meaning to accrue through the juxtaposition and 'unveiling' of chronotopes throughout the narrative.

Deserts and wastelands appear in Bradbury's, Clarke's and Heinlein's narratives and in later terraforming stories such as Frank Herbert's *Dune* (1965) and Ursula K. Le Guin's *The Dispossessed* (1999 [1974]). Clarke's and Heinlein's wasteland chronotopes, while sometimes inhibiting movement, are never as unremittingly stark as the landscapes presented by Miller and Anderson. The severity of their wastelands is mitigated by supportive social frameworks, in contrast to Miller's and Anderson's depiction of environments made relentless by repressive societies. Such physical landscapes often encode political meaning, with signifiers such as the planet's physical inhospitability and resource scarcity emphasised alongside a rigidly stratified and authoritarian society. A certain degree of secrecy and arbitrarily restricted movement accompanies these representations. The city, the labour camp and the wasteland deny exploration and enforce alienation in both natural and built environments. The claustrophobia of the social arrangements draws attention away from nature and onto political and individual aspects of the human experience on these worlds.

A satire on corporate America and the 'peculiarly American talent of advertising,' *The Space Merchants* uses the planned colonisation and terraforming of Venus to unveil a range of exploitative societal relationships that dominate the Earth of the text (Pohl and Kornbluth, 1974, 10). Mitchell Courtenay works at Fowler Schocken Associates as an advertising executive assigned to a campaign to recruit people for the project of colonising Venus. Colonisation becomes the site of a struggle between the ideological orientation of corporate America and that of the Conservationists ('Consies'), an underground organisation that opposes the almost total control that the corporations exert over the Americanised Earth. Courtenay is kidnapped, his death is faked, records of his identity are tampered with and he is contracted as a labourer to the Chlorella Corporation in Costa Rica. This Conservationist conspiracy is designed to remove Courtenay from his position as propagandist for the Venus project while simultaneously exposing him to the realities of life as a poor consumer/producer. Courtenay's struggle to regain his position exposes him to the extent of the Conservationist network and their plans to establish the Venus colony as an independent settlement founded on Conservationist principles.

Subordinated to the economic (and hence political) prerogatives of the corporate world, the American government in *The Space Merchants* possess severely limited powers. This arrangement anticipates the political engagement with multinationals in post-1970s terraforming stories such as Robinson's *Mars* trilogy (1996c [1992]; 1996b [1993]; 1996a). Fowler Schocken of Fowler Schocken Associates claims that '[w]e've actually and literally conquered the world. Like Alexander, we weep for new worlds to conquer' (Pohl and Kornbluth, 1974, 8), and he proposes the colonisation of Venus as an extension of their corporate mastery of Earth. This expansionist drive is a consequence of Schocken Associates' desire for social control through advertising, which is geared towards maximising profits via 'any act that served our god of Sales' (9). Illustrative of the power of Schocken Associates is 'Indiastries,' formerly India and now a cartel, an industrial factory that supplies all its produce to Schocken Associates for marketing. Venus represents a new industrial complex, a whole planet potentially available for financial speculation and exploitation. The Conservationists, popularly derided as a constituent of the 'lunatic fringe' (19), work in opposition to this hegemonic bloc. Their belief that 'modern civilization was in some way "plundering" our planet' and that 'Nature's way of living was the *right* way of living' contradicts the tenets of corporate America, thus prompting Schocken to dissociate the Venus project from them and to claim that 'spaceflight and Conservationism are diametrically opposed' (16–17).

The visuals for Courtenay's advertising campaign illustrate a fundamental dynamic central to terraforming: Courtenay observes how '[t]he airbrush and camera people were having fun sculpting a planet. It was the ultimate in "Before and After" advertising, and they were caught by the sense of history' (Pohl and Kornbluth, 1974, 55). This captures the essence of ideas of landscaping, focused here through a sensuous artistic metaphor. The technological adaptation of nature exhibits an autonomous artistic dimension that is driven by an impulse for improvement, but which is ultimately subordinated to economic desire.

This episode is juxtaposed with another scientific development pioneered by Schocken Associates' research and development team: one researcher notes that Venus has an abundant source of pure energy direct from the sun that can be tapped, a fact that stands in contrast to resource scarcity due to overpopulation on Earth. This direct source, unfiltered by distance, atmosphere and other factors, is a variant of the utopian dream of free energy offered by atomic sources in such works as Jack Williamson's CT stories. The transformative aspect of the 'Before and After' images of Venus is tied to a sense of the possible history that could develop on the planet: of change as a process constrained by natural factors but open to the potential for a willed transformation that would tie these natural processes to human agency. This sense of time pivots on the results of the actual terraformation of Venus, developments that are absent from the story but which are the subject of a struggle for determination throughout the narrative.

Although the plot of *The Space Merchants* ends with the arrival of the Conservationists on Venus and hence excludes occurrences of planetary adaptation (apart from the backgrounded terraforming of the Moon and the geoengineering of Earth), this text is important to the dystopian tradition of terraforming owing to its alignment of radical oppositional politics and Conservationism against a satirised corporate America. The ideological orientation of the corporations and the consumerism that they extol is identified as the cause of the regional effects that threaten the integrity of Earth's environments. Antarctica, a Schocken asset over which a bloody feud with their rivals Taunton Associates is fought, and Cal-Mex, a region encompassing California and Mexico that is subject to severe earthquakes caused by H-bomb testing, are environments affected by the primacy of corporate interests. The Conservationists and their organisation into cells recall fears of an underground communist organisation in America, yet their call for greater education and a general demand for 'planning of population, reforestation, soil-building, deurbanization, and an end to the wasteful production of gadgets' recalls the aims of the early energy economists in most respects bar the Conservationists' pastoral desire for deurbanisation (Pohl and Kornbluth, 1974, 101). The portrayal of the Conservationists as a revolutionary underground movement anticipates the Lunarians of Heinlein's *The Moon Is a Harsh Mistress* (2001 [1966]) and radical environmentalists such as the Reds in Robinson's *Mars* trilogy.

In 'Crucifixus Etiam,' Walter M. Miller depicts through the experience of the Peruvian colonist Manue Nanti the corporate exploitation of labourers involved in terraforming Mars. Nanti signs up for the terraforming project with dreams of earning enough to retire comfortably on Earth at a young age. Earth is a world of diverse landscapes alien to Nanti, who 'had seen so little of it that many of its places would be more alien to him than the homogenously ugly vistas of Mars' (Miller, 1973, 58). As a global chronotope, Earth is a series of natural and artificial spaces that nest, the whole signifying the plurality of the value systems and cultural voices on the planet. In contrast, Mars is a homogenous landscape, 'a nightmare, a grim, womanless, frigid, disinterestedly evil world' (61).

During an episode in which a Catholic priest from Earth visits the colonists to give Mass, Nanti witnesses him spill a drop of wine onto the Martian soil. This event functions as a pastoral emblem: the wine awakens Nanti's imagination to images of 'sunny Sicilian vineyards, trampled from the grapes by the bare stamping feet of children,' a landscape that he has only experienced imaginatively (Miller, 1973, 66). The red wine, Christ's blood but also 'the rich red blood of Earth, soaking slowly into the crust of another planet,' stands in for the blood and suffering of the colonists who are remaking Mars (66). If, as Nanti thinks, '[f]aith needed familiar surroundings, the props of culture,' then faith and culture are human expressions of a reciprocity between themselves and their environment, a feeling of being at home (67). On the homogenous landscapes of Mars, where the only environment that the labourers experience is a world of endless toil, faith in the project and in humanity cannot thrive and achieve expression through cultural forms adapted on Earth. 'Crucifixus Etiam' explores the possibility of developing a Martian culture that will allow the colonists not only to survive, but to eventually make a home of Mars.

Economic and cultural exploitation form the centre of the text's ecocritical engagement. This exploitation is primarily socio-economic, as the labourers are of diverse origin: Nanti's co-labourer is Tibetan while Sam Donnell, one of the veteran labourers known colloquially as 'troffies,' is presumably American. Nevertheless, Nanti's position as the protagonist raises the implication of a latent critique of cultural-economic exploitation in South America, and by extension the correspondence between ethnic and socio-economic exclusion. The supervisors who appear in the text, the German Vögeli and the (presumably American) Will Kinley, are of European and American descent and represent the perceived economic and cultural superiority of the 'West.' The relationship between labour and suffering is given emphasis by a visceral image of bodily adaptation that resonates with Bradbury's romantic treatment of Driscoll's physiological adaptation to Mars in 'The Green Morning,' itself a metaphor for psychological and cultural adaptation. Unlike the supervisors, who are quartered in environmentally contained habitats that afford them the luxury of living without breathing apparatus, the labourers are implanted with a system of hoses and valves that breathe for them, making natural, physical breathing obsolete. The impact of this form of pantropy on the labourers is suppressed: the outcome is the irreparable loss of the use of their lungs and an indefinite extension of their three-year contract. Labourers must reconcile themselves to exile, to an alienated existence with their own bodily otherness in an unwelcome environment.

Early in the story Nanti wonders, '[w]hy should men lose their lungs that after eight centuries of tomorrows, other men might breathe the air of Mars as the air of Earth?' (Miller, 1973, 75). When Nanti's motivation for contracting himself to the project can no longer be realised, disaffection leads him to seek new reasons for living. Kinley offers Nanti a pastoral aphorism, '[s]ome sow, others reap,' that turns on the theme of sacrifice and recalls the story's title (77). Nanti begins to see the project as 'an eight-century passion of human faith in the destiny of the race of Man' (78). This exportation of Catholic iconography onto Mars offers Nanti an alternative landscape as a means of coming to terms with his future in this new environment. He is able to bridge the discontinuity between Earth and Mars with a narrative of sacrifice and faith.

In contrast, Donnell offers an economic explanation for the terraforming project that echoes Jean-Baptiste Say's circular flow theory of economics: 'Mars is an outlet for surplus energies, manpower, money. Mars Project keeps money turning over, keeps everything turning over. [...] if the Project folded, surplus would pile up – big depression on Earth' (Miller, 1973, 69). This Orwellian economic vision sees terraforming as an extension of a capitalist strategy utilising the project as a check to economic collapse on Earth. Representation of economic expansionism draws from colonial traditions which, like the pastoral elements incorporated into the text's structure, offer spaces to critique civilisation's construction of new worlds on the foundations of the forgotten – on the marginalised and exploited members of society. As one of these forgotten, Nanti finds this explanation unsatisfyingly narrow and can only reconcile himself to life on Mars by entertaining a belief in a more expansive sense of purpose underlying the terraforming project.

Believing that 'if he were laboring for any cause at all, it was to build a world so unearthlike that he could not love it,' Nanti realises that the pastoral promise of a garden on Mars is fundamentally denied to him (Miller, 1973, 67). The difficulty of constructing a new home on Mars is compounded by social inequalities that ultimately proclaim the project to be oriented not towards himself and his peers but towards others who will reap the benefits of their labour. Nanti strikes another labourer during a crucial episode in which a spontaneous and limited rebellion threatens to occur and so negates the possibility of an uprising, but the subordination of his identity, desires and voice to a collective project does not stand unquestioned. An essential aspect of the story's design is the way promises of economic security are combined with pastoral imagery to structure a manipulative relationship between the directors of terraforming and those who actually work to transform the planet. Tragically, Nanti accepts his sacrificial role, but only after a psychological struggle with his bodily adaptations emphasises to the reader the severity of the exploitation visited on the colonist–labourers. Terraforming is used in this story, as in later stories of the 1950s, to expose the issue of the controversial foundations from which societies are built.

Poul Anderson's 'The Big Rain' and *The Snows of Ganymede* explore issues of nationalism and the contrast between democracy and communism, once again casting Earth as the political centre against which the terraformed planets are measured. Drawing on the Cold War paranoia prevalent in the 1950s to inform the dystopian elements of its Venusian society, 'The Big Rain' is a tale of oppression and espionage set after Venus's declaration of independence. Earth's relationship to the newly independent Venus is one of wariness: Hollister thinks that 'however peaceful Earth might be, she was still a shining temptation to the rest of the system' (Anderson, 2001, 5). The story insists on an interplanetary connection in the form of a United Nations, an expansion of the global sense of social co-ordination to interplanetary scales. While Venus is far from united, the text emphasises its status as a global state and questions its relationship to the interplanetary confederation, of which Earth is the central political body. Again Hollister makes explicit that this political clash can be seen in terms of the clash between democracy and communism on the one hand and the regulation of groups and the individual on the other. The latter in both instances leads to an unhealthy Cold War paranoia. This anxiety underlies the relationship of the colonies to their centre and highlights an instability underpinning the UN-affiliated interplanetary society.

As in *Farmer in the Sky* and *The Sands of Mars*, the harsh landscapes of the planets to be terraformed may justify a certain degree of individual subordination to the group, but, Hollister argues, 'that doesn't give anyone the right to collectivize the minds of men' (Anderson, 2001, 31).

Regarding the fact of rationing in the capital New America, Hollister believes that Venus's harsh conditions promote a communist ethic: 'He reflected that the communist countries before World War Three had never gone this far. Here, everything was government property. The system didn't call itself communism, naturally, but it was, and probably there was no choice. Private enterprise demanded a fairly large economic surplus, which simply did not exist on Venus' (Anderson, 2001, 6). As in 'Crucifixus Etiam,' the colonists on Venus exist for the sake of production; their lives have become subordinated to the distant goal of a terraformed planet. This drive towards material production is used by the government to legitimate tools of state control. Those deemed by the secret police to be enemies of the state are exiled to the mines at Lucifer, where, despite Hollister's argument that '[y]ou could use free men, taking proper precautions, and it would be a lot more efficient and economical of manpower,' they are callously overworked and habitually exposed to radiation (Anderson, 2001, 26). Lucifer's main function as a labour camp is as a deterrent and detention centre for malcontents.

Venus's landscape is accorded only instrumental value, leading to its terragouging in the narrative's past. Hollister's first supervisor, Heinrich Gebhardt, sees an exploitative dynamic at work in Earth's initial imperialist mining of Venus when he explains that 'the so-called democracies often relied on broken men, who could not find vork at home or who had been displaced by var. No, ve owe them notting' (Anderson, 2001, 5). The terms of Gebhardt's rejection of Earth inform Venus's independence and speak of a desire to escape the exploitative relationships that an imperialist colonisation of the solar system entails. The distances involved and the prohibitive cost of space travel add a further dimension to this political relationship, as

> a military expedition to suppress the nationalists would cost more than anyone could hope to gain even from the crudest imperialism. Also, as long as no clear danger was known to exist, it wouldn't have sat well with a planet sick of war; the dissension produced might well have torn the young world government, which still had only limited powers, apart. (Anderson, 2001, 6)

It is not on ethical grounds that the suppression of nationalist independence is rejected but on economic, logistic and political grounds. Secret policeman Captain Karsov explains that one reason for the schism between Earth and Venus, and the latter's refusal to remain a part of the UN, is that 'we are ... necessarily ... developing a whole new civilization here, something altogether remote from anything Earth has ever seen'

(Anderson, 2001, 5). Hollister believes that this attempt to establish an independent culture grounded in nationalism is a propagandistic aid to the systematic oppression of the majority of the colonists. Nevertheless, it exposes a reaction to the political asymmetry between Earth as the colonising centre of an interplanetary government and its peripheral colonies.

This story takes its name from the Big Rain model of terraforming, the metaphorical potential of which has been returned to in several later terraforming narratives (during the conclusion of David Lynch's cinematic adaptation of *Dune* in 1984, for example). In 'The Big Rain,' colonists distribute seven million 'airmakers' over a period of thirty years across Venus's surface. Combined with the manufacture of soil and a series of strategically deployed hydrogen bombs designed to initiate volcanism on Venus, this project, it is thought, will release enough water into the atmosphere to cause a global downpour lasting ten Earth years, at the end of which

> there would be rivers and lakes and seas on a planet which had never known them. And the soil would be spread, the bacteria and plants and small animal life released. Venus would still be mostly desert, the rains would slacken off but remain heavy for centuries, but men could walk unclothed on this world and they could piece by piece make the desert green [...] In five hundred years, all of Venus might be Paradise. (Anderson, 2001, 13)

This pastoral image compresses the temporal scope of the intermediary stages involved in bringing life to other planets as much as it gestures towards the idea that satisfactory initial conditions form a precondition to future ecologic and spiritual nurture and growth. The physical attributes of other worlds are used to offer an ecological image whose dynamism exceeds the scale of the natural phenomenon of Earth. Through this image the sentimental kind of pastoral theme enters into the text as a politicised form of appeasement for the exploited workers. The event of the Big Rain and the vision of a far-future paradise is the subject of humorous songs that cast this future as a land of Cockaigne (10). An ecology that is geared towards human needs and consumption sits behind this depiction of a land of fantastic abundance where earthly desires are fulfilled. As in Miller's text, however, this landscape is displaced into an inaccessible far future. Indeed, the pioneering spirit involved in romantic representations of terraforming has no place on Venus: Karsov warns Hollister that '"Venus is no place for the rugged individualist [...]. Men have to work together, and be very tolerant of each other, if they are to survive at all"' (Anderson, 2001, 3). What the text leaves us with is not a community driven by a pioneer spirit but a society in which the sentimentalised pastoral is variously co-opted or denied in order to sustain an asymmetric socio-political arrangement designed to maintain the productivity of a workforce in the face of extreme environmental conditions and exploitative social relations.

Anderson's *The Snows of Ganymede* takes place in the same Psychotechnic universe as 'The Big Rain.' In the far future, nationalist political bodies are replaced by global organisations such as the monastic Order of Planetary Engineers, whose relationship to Ganymede parallels that of the earlier UN to Venus in 'The Big Rain.' The pastoral pioneer tradition is revisited in *The Snows of Ganymede*, in which the history of the initial colonisation of Ganymede by the White American Church repeals the Pilgrim Church's earlier colonisation of Mars (Anderson, 1958, 15–19). The political opposition between the centre and the periphery of the interplanetary society defines a contrast between the oligarchic state descended from this religious organisation and the Order, headquartered on the Moon. The Ganymedeans contract the Order to adapt Callisto's environment; like Hollister in 'The Big Rain,' this allows aspects of the Jovian government to be unveiled in the course of the Order's presence on Jupiter's moon. Their investment in planetary engineering is based on a utilitarian ethic combined with a militaristic pride and spirit, ideal characteristics providing moral foundations for the Order's mission 'to make space available for all men, regardless of race, creed, or political affiliations' (15). The Order denies political involvement of any kind; their motivations are supposedly practical and technical, and their understanding that various methods of planetary adaptation are necessary to cope with the idiosyncrasies of global environments, 'that other worlds were not Earth,' suggests a corresponding liberalism towards political groups (22). However, this motivation is coupled to one that speaks directly to their power as an organisation: the project will give them the monopoly they need to ensure their safety in the interplanetary order (30).

This monopoly seems to contradict the Order's apolitical stance, and the Engineers find they must grapple with questions of power and a possible disengagement from politics. On a narrative level these issues are explored and resolved through the consciousness of the protagonist Hall Davenent. Having only ever faced 'the inanimate savagery of planets,' Davenent has been shielded from the way in which encounters with human hostility call forth a political response (Anderson, 1958, 34). The Order's aim, 'to keep the scientific spirit alive. To reform planets, not people,' is central to this issue (52). The Psychotechnics – practitioners of a science of directed social evolution – have developed a range of practices aimed at the reformation of cultures, and so stand directly opposed to the Order. The insight that environments influence cultural development, 'that man necessarily develops a different civilization in every environment if he stays long enough, and that what may shock you is normal, perhaps necessary, on Ganymede,' buttresses the relationship between the environment and socio-political orientations embodied by the terraforming motif (16–17). Davenant ultimately chooses political intervention when he orchestrates a rebellion and forces a new government upon the colonists.

Intellectual isolationism and social oppression are the core political issues that Anderson explores, and he links dogmatic religion to the dystopian government as a symbol for this isolationism. Enlightenment science, aligned with the spatial engineers, is the other side of the duality that underlies the political opposition between Earth and Ganymede. The conflict between Ganymedean society and a group of outlaws defines another intersection between politics and the pastoral. This outlaw community lives at the fringes of the Ganymedean society and prefigures similar uses of such groups in *Dune* (the Fremen) and the *Mars* trilogy (the Reds and Zygote). Unlike these later manifestations, the outlaws in *The Snows of Ganymede* have not been able to develop a sustainable relationship with their environment, resulting in a deterioration of their living conditions and an existence as 'barbarians' (Anderson, 1958, 72): '[t]he psycho-social effect of alien conditions had yet to be measured. Huddling, hiding, waging a doomed war for three or four generations, the hill men would rapidly have forgotten their intricate, highly specialized civilisation. The barrenness and cold of the landscape would have entered their souls' (Anderson, 1958, 74). The narrator exhorts the reader to '[b]ehold the nobel [sic] savage!' when Davenent and Kruse, the last two surviving members of the terraforming expedition, witness the results of this social decline (Anderson, 1958, 77). The outlaws are thus ironically likened to a pastoral, backward-looking society, in contrast to the specialised technologies and science that ground civilisation's ability to improve humankind's living conditions. Scientific measurement enables the reconstruction of social trends, placing the Engineers as scientists in a position of political agency that justifies (for them) their use of the outlaws as a revolutionary force to establish a new, rational government on Ganymede. Recalling Hollister's empty promises on behalf of the UN to the rebels of Venus in 'The Big Rain,' Davenant makes compromises with the terraformers' charter and with ideal modes for establishing new governments and political relations between members of the interplanetary society.

Moral Extensionism in Terraforming Stories of the Late 1950s and Early 1960s

The American sf that deals with terraforming themes in the 1940s and early 1950s develops some of the ideas Wells and Stapledon raise in their 1930s scientific romances. However, these pulp stories foreground perspectives limited by single characters or groups, and often avoid temporal juxtapositions of a multiplicity of different societies and their attitudes to space. Thus the expansive future history and essays in myth creation that characterise Wells's and Stapledon's use of terraforming are backgrounded in favour of the synchronic juxtaposition of multiple landscapes. Anderson's 'Call Me Joe' (1989 [1957]) uses this dynamic to contrast the sterile spaces of Earth to the wild beauty of a harsh Jovian landscape; between these spaces a monastic scientific research station acts as a middle landscape that mediates between the two. Nevertheless, in the example of Anderson's Psychotechnic series, such broad sweeps of time are generated via an intertextual, dialogic element in which multiple stories portray landmark episodes in the history of an imagined interplanetary society for speculation and commentary on political themes. A temporal juxtaposition between the reader's contemporaneity and the text's present, past and future resonates with the structure of the future history.

Heinlein's *Farmer in the Sky* and Clarke's *The Sands of Mars* include episodes in the latter parts of their narratives that introduce the discovery of alien life into their stories. Terraforming in the 1950s reflects ideas that were being developed in a wide range of contexts, but shows a distinct engagement with the politics of imperialism, nationalism and utopia. When these themes intersect with the trope of alien life, terraforming is often used to explore the extension of human moral environments towards others, as analogues either for terrestrial flora and fauna or for societies. The latter category is, in an important sense, also nature's otherness: humanity is a part of the all-encompassing category of cosmological nature. The intelligent alien or the alien society (scalar levels) offer spaces for reflecting both on nature's (alien) otherness and on cultural otherness. While many alien societies are not other precisely because they are direct analogues for Earthbound societies, many terraforming texts open up ethical debate on issues of responsibility and respect for nature's otherness by depicting encounters with aliens whose possible intelligence remains unresolved.

The ethics of colonisation are not straightforwardly consistent with environmental ethics. The extension of culture-bound ethics risks subordinating the needs and interests of others to those of the self.

As Stapledon and Bradbury illustrate, colonising planets inhabited by alien civilisations raises related engagements with intra-human ethical responsibilities. When these alien civilisations are taken as signifiers of nature's otherness and their cultural status is bracketed out, an environmental ethics that reflects on nature's otherness can be brought into contact with the cosmological speculation of the text. These two separate ethical domains, or ways of framing ethical consideration, support each other: moral extensionism highlights similarities between others; nature's otherness emphasises difference. Alien intelligences are often accorded intrinsic value based on their perceived degree of sentience and intelligence. Non-sentient, non-human others are often accorded less or no intrinsic value, while abiotic nature tends to be seen as valuable for extrinsic reasons only: for its provision of the environment necessary for biotic life to thrive.

Anderson's 'Sister Planet' (1960 [1959]) uses the theme of human–alien relations to reflect on extermination as part of the terraforming process. This idea received more attention in terraforming stories throughout the 1960s, responding in part to the developing sense of urgency that also inspired the American environmental movement. Arthur C. Clarke's 'Before Eden' (2001 [1961]) is a story of environmental contamination (and in retrospect an unintentional ecopoiesis) that broaches themes dealt with in Carson's *Silent Spring*, serialised a year later in the *New Yorker*. While alien life has been correlated with terraforming narratives since the 1930s, an investigation of short stories from the late 1950s to the early 1960s highlights an increased interest in issues of moral extensionism that inform one facet of the use of the terraforming motif in sf. These engagements, as with the utopian and dystopian traditions discussed above, make use of pastoral elements and techniques to frame their exploration of the ethics of colonisation.

'Sister Planet' tells of Nat Hawthorne's return to the research station on the ocean planet Venus, where, as a biologist, he studies the ecology of its indigenous life. The research station is supported solely by a trade in firegems, which appear to be a biological product of the dolphin-like cetoid life forms dominating the Venusian seas. Despite human–alien trading relationships with the cetoids, their intelligence continues to be questioned. Hawthorne's close relationship with a cetoid called Oscar encourages him to credit the species with intelligence, and when Oscar takes him to view one of the cetoids' undiscovered underwater 'cities', he becomes convinced that they have developed an advanced culture fundamentally alien to humanity. Dazed by this experience, he withholds his discovery from the scientific community on Venus until the geophysicist Wym Dykstra unveils his analysis of the Venusian core and proposes 'the largest and most significant engineering project of history. [...] The colonization of Venus' (Anderson, 1960, 114–15). Dykstra's proposal entails the cetoids' extinction, encouraging Hawthorne to protest and announce as evidence for the cetoids' intelligence his trip to their sunken city. As a result, Dykstra and Captain Jevons agree that terraforming Venus would be unacceptable, and they resolve to bury the results of Dykstra's research, because Earth, pressured by overpopulation, would condone their extinction. Hawthorne is dissatisfied with this development and orchestrates a plan to prevent future research from discovering Venus's suitability for terraforming. He bombs the station, killing the rest of the community, and, in an echo of the genocide of the Venusians in Olaf Stapledon's *Last and First Men* (1966 [1930]), indiscriminately kills many cetoids in an assault designed to sow distrust between humanity and the aliens. With the lucrative trade in firegems sabotaged, there is no reason for Earth to continue developing interests there, thus preserving the planet from terraformation.

The scientific community's scepticism about cetoid intelligence is grounded in its failure to achieve cross-species communication, the cetoids' 'childlike' attributes (playfulness and innocence), their disinterestedness in books as trade items and the scientists' belief that '[i] ntelligence is supposed to evolve in response to a rapidly changing environment' (Anderson, 1960, 104). These reasons are based on anthropomorphic assumptions that cast the aliens in the role of primitive, pastoral creatures, while the last reason, an evolutionary notion, relies on geographic and climatic knowledge of Venus that is incomplete. Hawthorne responds to the cetoids with a mixture of acknowledgement (of their otherness) and identification; he argues that communication has not yet been established because 'our minds are too strange' (101). This sense of alien otherness is reinforced when Hawthorne sees the cetoids' underwater structure and observes 'patterns [which] were so strange that his mind was not trained to register them' (107). By assuming the fact of the cetoids' consciousness, Hawthorne is able to recognise their agency and intelligence as a basis for an identification with the other. This recognition invites a form of transgredience that allows Hawthorne to imaginatively assume an external evaluative position with regard to the human relationship to the cetoids and Venus. He is thus able to conceive a metaphorical sense of kinship with the aliens, leading him to extend the range of his familial relationships to them by greeting and recognising them in human terms: '[w]elcome, my brother' (105).

This moral extensionism does not constitute an environmental ethic that can account for an appropriate human relationship to nature's otherness in Simon Hailwood's terms, where nature is valued extrinsically and relationally by virtue of its otherness. What is outlined instead is a strong version of respect for consciousness (intelligence) and not non-human nature as such. Hawthorne's metaphorical extension of kinship categories to encompass cetoids as brothers disrespects their otherness, but is balanced against his recognition of their irreducible difference. Nevertheless, Dykstra's terraforming proposal shocks him because it threatens creatures who, as intelligent beings, deserve to be factored into instrumental decisions involving their extermination for anthropocentric ends. Non-sentient biotic and abiotic nature is valuable only insofar as it provides the background for the cetoids' existence. The relationship between the cetoids and the human settlement combines the political opposition between centre and pastoral periphery with issues of environmental ethics.

Hawthorne's decision to sabotage the human–alien trading relationship is a prime example of Fogg's notion that arguments against terraforming often embody misanthropic critiques. Hawthorne's position could be called 'misanthropic despair': he reflects that *'I don't care very much about humankind. It's Oscar I want to save. And how much hate for one race can hide under love for another?'* (Anderson, 1960, 125). This despair is grounded in an outlook that operates by a metaphorical extension that correlates scientific, aesthetic and ethical values. Hawthorne is somewhat unfairly able to claim that 'we've created one beautiful thing with all our ingenuity – just one, space travel. I'm not sure how much destruction and ugliness that makes up for' (99). Jevons contradicts Hawthorne, suggesting that not only art, but 'the beauty of science itself' points towards redeeming human qualities and hope for the future (99). A contrast is made between Earth as an emblem for socio-political problems, exacerbated by overpopulation and contextualised against an awareness of historical oppression, and the Venusian planet as an aesthetically rich symbol of nature populated by an innocent, pastoral civilisation exempt from the corruption of Earth.

The ocean planet is a place where 'all that you saw would be beautiful' (Anderson, 1960, 89). The research station is located in Venus's Phosphor Sea, so named for its bioluminescence, which signifier cues a response to beauty. The cetoid city is a locus for this luminous aesthetic; Hawthorne thinks of it as 'a ragged jumble of spires, bluffs, and grottos, eerie but unorganized beauty,' '[a] city of merfolk,' which is constructed of 'arches and buttresses of fragile filigree, [possessing] an overall unity of pattern.' He is unable to decide whether it is 'a memorial. An art gallery, or – Hawthorne didn't know' (107). These structures signify the city's physical fragility and tap into the Burkean notion that delicacy and softness, both physical and visual (which bioluminescence also conforms to), engage a response to beauty in the perceiver that is grounded in the object's submission to the subject, in this case the cetoids to Hawthorne's exploratory gaze.

Before Hawthorne sees the city he entertains the notion that, *'by sheer telepathy or something, they* [the cetoids] *build their messages into the crystal structure of stones on the ocean bed'* (Anderson, 1960, 104). This impression makes literal the notion that intellectual landscaping assigns human meaning to the environment. Hawthorne's aesthetic response to the city is therefore directly linked to an ethical appreciation of the cetoids as intelligent beings and underlies his speculation that they perhaps possess 'more soul – more sense of beauty and mercy and laughter – if you extrapolated their present behaviour' (105). In the light of his visit to the city, Hawthorne concludes that the cetoids are intelligent, not because they are able to build structures that suggest an advanced civilisation, but because 'the contemplation of beauty is essential to thinking life. [...] The underwater blending of so much that was constructively beautiful could not be a freak of nature' (Anderson, 1960, 108). Aesthetic responses to nature and art are necessary conditions for the development of intelligence in Hawthorne's view. That he is able to respond aesthetically to the cetoids' landscaping suggests a degree of commonality between human and cetoid intelligence. Just how far does Hawthorne's attribution of beauty reflect his own landscaping of Venus and the cetoids? He has already called the cetoid structure a 'city of merfolk,' landscaping both the environment and its inhabitants, and he later settles on the reference 'the holy place,' which suggests that he is unable to break away from his own intellectual landscapes (Anderson, 1960, 113). Nevertheless, there is room for the recognition of nature's otherness within these spaces, illustrating how the inevitable process of intellectual landscaping and the recognition of otherness can be overlaid onto the same object, thus satisfying Hailwood's condition that nature be respected for its otherness and that particular meanings landscaped onto any space should not be confused with all its meaning. Hawthorne quite clearly demonstrates this when he thinks that '[h] uman senses and human science didn't exhaust all the information in the cosmos' (102). In this context the pastoral operates as one of these landscapes and structures the cetoids' relationship to the humans according to a traditional country/city opposition. Anderson combines this element of pastoral structure with sf techniques analogous to the pastoral's compression of meaning to comment on the instrumentalism of an imperial attitude to space, but overlays this with a critique of misanthropy embodied by Hawthorne's mass slaughter of the cetoids. Hawthorne's scheme removes autonomy from them as much as it does from humankind, substituting instead the monologia of an individual's determination of future human–cetoid relations.

Clarke's 'Before Eden' relates an expedition's discovery and unwitting destruction of alien life. Ecopoiesis of inhabited planets can be seen as an infection; 'Before Eden' is ultimately a parable of contamination. The narrator concludes the narrative by recounting an alien organism's consumption of the refuse left by human visitors on Venus, from which it *'absorbed a whole microcosmos of living creatures,'* a few of which manage to survive and infect the alien; as in *The Martian Chronicles*, *'it carried contagion to all its world.'* The biblical voice of Genesis is bitterly evoked: *'[b]eneath the clouds of Venus, the story of Creation was ended'* (Clarke, 2001, 68). The text raises ideas consonant with environmental stewardship, the belief that humanity could act in a benevolent and wise capacity to preserve aspects of nature in the case of aliens that cannot communicate through language, and yet 'Before Eden' ends by undermining humanity's sensitivity to their own harmful acts. Such notions underline the hubris latent in environmental stewardship and implicitly subvert God's concession to humanity of dominion over life on Earth. The emphasis on the sublimity and wonder of the alien contributes to the tragedy of the loss of the other, as does the loss to the scientific domain of a rare object of study. The naive vision of life's call illustrates the dangers of confusing nature's otherness with a teleological conception of cosmological nature oriented towards humanity's own existential shortcomings. The Edenic paradise that terraforming promises in many of the consensus futures of the 1950s is in this story the death knell of an idyll formerly exempt from humankind's hubris and corruption.

Nature's alien otherness is valued extrinsically and on aesthetic grounds through the perceiving humans' relationship to the life form:

> The transformation was so stunning that neither man could check a cry of astonishment. Gone in a flash was the deep, somber black of the thick-piled velvet carpet at their feet. Instead, as far as their lights carried, lay a blazing pattern of glorious, vivid reds, laced with streaks of gold. No Persian prince could ever have commanded so opulent a tapestry from his weavers, yet this was the accidental product of biological forces. Indeed, until they had switched on their floods, these superb colors had not even existed, and they would vanish once more when the alien light of Earth ceased to conjure them into being. (Clarke, 2001, 66)

Description of the alien evokes a sense of the Burkean sublime, which highlights obscurity and privation as governing factors of nature's otherness ('somber black'). Sudden transition as productive of the sublime, bright colouring and astonishment are also evident in the description of the scientific expedition's first encounter with the alien. The reference to Persia acts as a signifier of opulence and otherness, in this case a cultural otherness that stands in for the alien otherness of the Venusian, thus alluding to a problematic link between aliens as natural and cultural others and issues of moral extensionism. This image also contrasts and de-privileges the artificial in favour of the natural. The aesthetic superiority of nature as biology over art is emphasised, as is the fact that recognition of such beauty is dependent on the presence of an alien force (flood lights) and on intelligent (human) observers capable of perceiving this beauty. Their embodied perspective provides the necessary conditions for an aesthetic response to nature.

The text offers other reasons for valuing nature's otherness. The alien does not display any sign of consciousness, let alone intelligence, thus sweeping aside arguments that ethical responsibilities are owed only to intelligent beings. The scientific interest of Hutchings offers another reason for valuing aliens extrinsically, on the grounds that they present rare opportunities for scientific discovery. At the narrative's beginning he argues strongly for the possibility of life on Venus, pointing out that '[w]here there's water, there may be life' (Clarke, 2001, 61), and that, even in boiling water, '[t]here are algae that manage it [survival] on earth. And if we've learned one thing since we started exploring the planets, it's this: wherever life has the slightest chance of surviving, you'll find it' (Clarke, 2001, 62). Lessons derived from knowledge of Earth's extremophiles are extended towards alien environments as arguments trading on science's role as an explanatory tool for understanding the universe and for observing affinities between life forms throughout the cosmos. When the explorers to Venus eventually discover alien life, the narrator suggests an extrinsic value based on a metaphysical conception of nature oriented towards humanity:

> For life called to life, across the gulfs of space. Everything that grew or moved upon the face of any planet was a portent, a promise that Man was not alone in this Universe of blazing suns and swirling nebulae. If as yet he had found no companions with whom he could speak, that was only to be expected, for the light-years and the ages still stretched before him, waiting to be explored. Meanwhile, he must guard and cherish the life he found, whether it be upon Earth or Mars or Venus. (Clarke, 2001, 67)

Because this call of life is projected beyond human environments, it works as a justification for space exploration. This is compounded by another motive, the search for proof that human beings are not alone in the universe. This existential question bears on the theme of the desire for meaning within an indifferent universe, the ultimate aim of which is to achieve communion with other intelligences. This understanding of science acts as a form of landscaping in which nature's otherness is incorporated into human cultural landscapes as a 'promise.' Implicit here is that biotic life is viewed teleologically, as part of a vision of the universe that privileges life precisely because it is rare, and hence abiotic nature's otherness is backgrounded.

This chapter began by considering how early 1950s terraforming stories were informed by pastoral themes and deployed techniques analogous to the Elizabethan pastoral. The visions of the future conceived in these works developed in a dialogic relationship that helped shape the terraforming megatext. These narratives relied heavily on the American pastoral, which was used to frame terraforming in terms of the colonisation of America and which allowed writers to highlight the incongruities between pioneer landscapes and the reality of space colonisation. As the terraforming tradition began to shift from optimistic, utopian visions of interplanetary colonisation and terraforming to pessimistic reflections on the dystopian aspects of interplanetary isolationism, a space for the exploration of political and economic oppression was established. These dystopian visions reflect on the failure of terraforming to realise the pastoral dream of a return to close-knit communities in the face of what was perceived as an increasingly alienating and global-scale expansion of space.

This chapter then returned to Anderson and Clarke to connect these political issues to environmental ethics in two works of the late 1950s to early 1960s. Terraforming as a form of landscaping is a way of coping with the asymmetry of human relationships to nature and other human groups. The physical fragility of the natural environment is complemented by an awareness of the fragility of civilisation, as Bradbury shows, but this is also paralleled by an intellectual fragility that makes it difficult to keep a respect for nature's otherness in view. Disrespect of nature's otherness often involves a parallel disrespect of human otherness, evidence of a structural homology between the attitudes of the agent of terraforming towards cultural and natural otherness. The imposition of anthropocentric needs and interests onto nature functions as a precursor to narrowly instrumental relationships. While nature's otherness cannot be directly portrayed in these texts, representation of the human response to multiple environments and to instances of alien life offers vectors for considering how landscaping the environment takes into account the needs and interests of society and nature. Underlying these uses of the terraforming motif are pastoral emblems that come to be identified as essential megatextual tropes for the terraforming narratives of later years.

3: Ecology and Environmental Awareness in 1960s–1970s Terraforming Stories

The 1960s–1970s New Wave of sf was a period of experimentation during which the tropes of the 1950s pulp tradition were transformed. Ecological concerns became an integral element of the socio-political engagement of the terraforming stories of this period. Responding to the impact of the 1960s counterculture, sf began to diverge from the consensus futures of the 1950s. Depictions of proto-Gaian worlds were directly associated with the motif of terraforming and presented counter-arguments against planetary adaptation. Although ecological themes have fed into representations of terraforming since the 1930s, terraforming stories of the 1960s–1970s were premised on the emergence of new ecological landscapes that reflected a sense of environmental urgency. The scientific aspects of ecology were combined with a strand of mysticism reminiscent of early twentieth-century vitalism, but were informed by a growing popular interest in socio-political institutions and practices alternative to those ideologically dominant at the time. This shift owes its character to the proliferation of ecologically oriented communes and the growth of 'hippie' culture in America, along with experimentation in ecological design, appropriate technologies and sustainability, which were widely popularised by Stuart Brand's *Whole Earth Catalog* (1968). These countercultural influences provided new contexts for the terraforming narrative to respond to the ways sf had traditionally landscaped nature.

Anna Bramwell argues that the counterculture of the 1960s was preceded by a distinctive tradition of radicalism that had existed for 150 years: 'not only had Utopian communities been founded by native Americans and recent European immigrants, but a strain of Asiatic mysticism and occultism among communards had an equally long history' (1990, 94). If a commune is, as Bramwell argues, a planned social experiment, then the colonies of the 1950s terraforming narratives are also communes. The experimentation involved in transforming the environment is linked to a social experiment with motives that, while certainly ecological, were also influenced by religion and politics (92). The Findhorn Foundation, established in Scotland in 1962, exemplifies the new ecological discourse that was being cultivated in the 1960s. Promulgating a belief in a spirit world underlying the physical and the coming of a 'New Age' 'that would be born when men realised their place in nature,' such communities re-contextualised the experiment underlying the commune and influenced a multitude of other environmental discourses, which in turn influenced sf (100–01). Brian Stableford has argued that the mystical aspects of ecology more often than the scientific have 'forged a crucial bond with the history of utopian thought, helping to redefine notions of eutopia (and hence of dystopia) and eventually necessitating the coinage of the term "ecotopia"' (2010, 259). Ecology offered an all-embracing paradigm that could cater to what Tom Moylan has, in the context of the critical utopia, referred to as an alliance of independent groups and interests converging on autonomy; in ecological terms this autonomy is predicated on an awareness of the interconnectedness between elements of nature (1986, 27–28). Links between the communes of the 1960s–1970s counterculture and the utopian and dystopian communities depicted in terraforming narratives since the 1950s illustrate a complex weaving of multiple discourses within the sf megatext.

In contrast to this tradition of mysticism was another that sought solutions to environmental problems through innovative ecological thinking and design, the appropriate use of technologies and environmental advocacy. This approach was pragmatic and optimistic in orientation, and promoted conscientious use of environmental research and science to explore alternatives to contemporary practices and to effect wide-ranging change through the mobilisation of individuals who would take the shaping of their destiny into their own hands. Stuart Brand's *Whole Earth Catalog* was the voice of this tradition, and its popularity inspired many to explore ways to engage with environmental issues. The *Whole Earth Catalog* collects a range of articles from reviews of books and technologies, surveys of the work of various thinkers in a range of disciplines, historical primers and other items, all of which are subordinated to the subtitle of the text, 'access to tools.' Among articles on Buckminster Fuller and tensile structures, the first issue reviewed Steve Baer's *Dome Cookbook* (1967) and the geoengineering text *Man's Role in Changing the Face of the Earth* (Thomas, 1971 [1956]), along with Frank Herbert's *Dune*. In the 1970s–1980s writers such as William S. Burroughs, Ursula K. Le Guin and Ernest Callenbach contributed to the *Whole Earth Catalog* and its successors and revivals, along with scientists James Lovelock and Lynn Margulis. Brand controversially explored the issue of space colonisation and produced a collection on the theme entitled *Space Colonies* (1977), which was influenced by the ideas of Gerard O'Neill and which emphasised the link between sf and the pragmatic, technologically oriented environmentalism of the *Whole Earth Catalog*. Although Andrew G. Kirk notes that the *Whole Earth Catalog* is informed by an environmental utopianism that was especially clear in the debate over space colonies, he argues that '[t]rends that were initially utopian often get tempered by time, evolving into more practical versions of the revolutionary thinking that spawned a period of great creativity' (2007, 10). Sf influenced the revolutionary thinking Kirk describes; the dialogism of the mode can be seen as a feedback system that refigures tropes and narratives in response to the needs and desires of society. Like the *Whole Earth Catalog*, the terraforming tradition can be read as a catalogue of tools for adaptation and as responses to landscaping and technologically driven societal change.

It was not until James Lovelock's popularisation of his hypothesis during the 1970s–1980s that the Gaia motif became a symbol for the blend of mysticism and science in ecological sf. Nevertheless, stories such as Richard McKenna's 'Hunter, Come Home' (2001 [1963]) and James White's 'Major Operation' (1971a) depicted the terraforming of proto-Gaian worlds, thus anticipating the important links between these two motifs. The Gaia hypothesis originated from Lovelock's work as an independent scientist at NASA in 1965, but was first widely popularised in his 1979 *Gaia* (1987). This hypothesis 'views the biosphere as an active, adaptive control system able to maintain the Earth in homeostasis' (Lovelock, 2006, 22). According to Lovelock, living organisms actively influence their environment in ways that stabilise climatic fluctuations and help maintain a state of regulation. Lovelock argues that Earth's apparently discrete ecosystems are connected and form a vast planetary system that can be thought of as a single superorganism. Like a bee's nest, superorganisms are 'bounded systems made up partly from living organisms and partly from nonliving structural material' (Lovelock, 1995, 15). This global outlook considers both biotic and abiotic processes as parts of a system, necessitating a holist approach to scientific practice.

This holism was informed by theories of cybernetics first popularised by Norbert Wiener's foundational work, *Cybernetics* (1948), and by his examination of the social aspects of cybernetics in *The Human Use of Human Beings* (1950). Concepts such as feedback systems, communication and control were central and greatly influenced the whole systems theory that provided a basis for the ecologism of the *Whole Earth Catalog*. Brand was inspired by Buckminster Fuller and his notion of postscarcity and

Spaceship Earth, his 'dymaxion' design philosophy, which promoted efficiency and called for multidisciplinary or omnidisciplinary thinking, his invention of the geodesic dome and his creation of the World Game, which challenged players to find solutions to ecological and social problems for the whole of the global population. In 1975, Lovelock and his collaborator Lynn Margulis published an article in the *Whole Earth Catalog*'s successor, the *CoEvolution Quarterly*, which disseminated the Gaia hypothesis to a wider American readership already primed by the exploration of related ideas in ecological design and cybernetics.

In 1969 Lovelock's neighbour William Golding proposed 'Gaia' for the name of his scientific hypothesis. This suggestion controversially associated the Greek goddess Gaia, a personification of the Earth and related to the pagan image of a Great Mother, to a scientific hypothesis which implied that the Earth itself was like a living organism. This was compounded by Lovelock's own spiritual outlook: he claims in the chapter 'Gaia and God' in *The Ages of Gaia* that '[l]iving itself is a religious experience' (1995, 192). Although he is adamant that his view of Gaia is not religious, that Earth is not analogous to a god and possesses no consciousness of its own, he explains that the fact that 'Gaia can be both spiritual and scientific is, for me, deeply satisfying' (204). New Age spiritualism greatly developed the implications of this mythic image of a Great Mother in Lovelock's popularisation of the Gaia hypothesis, and yet the Californian ecologism promoted by the *Whole Earth Catalog* also emphasised a synthesis of nature and technology that rejected spiritualism and the collective, technocratic management of the environment in favour of a pragmatic enthusiasm for alternative, low-impact technologies.

The terraforming texts of the 1960s–1970s can be divided into those narratives that present a clear continuity with the stories of the 1950s in terms of their form, themes, and narrative elements and the proto-Gaian narratives distinctive to this period. A third trend appears towards the end of the 1970s and develops the implications of pantropy in relation to terraforming. Including such stories as John Varley's 'Retrograde Summer' (2001 [1975]), Frederik Pohl's *Man Plus* (2000 [1976]) and David Gerrold's *Moonstar Odyssey* (1977), this trend will be excluded from consideration for reasons of space and because the works would be more productively considered in the context of a wider discussion of transhumanism, posthumanism and cyberpunk. John Brunner's *Stand on Zanzibar* (1999 [1968]) and *The Sheep Look Up* (1972) have made significant contributions to ecological sf. His less well-known novels *Bedlam Planet* (1982 [1968]) and *The Dramaturges of Yan* (1974 [1972]) explore terraforming and Gaian themes, yet they would also be better examined in the context of human–animal studies and posthumanism. One group of narratives in this period engage in dialogue with the earlier 1950s tradition of terraforming stories and look forward to those of the 1980s–1990s. Stories belonging to this group include Frank Herbert's first three *Dune* novels (1965; 1971 [1969]; 1976), Robert Heinlein's *The Moon Is a Harsh Mistress* (2001 [1966]) and Ursula K. Le Guin's *The Dispossessed* (1999 [1974]).

Another group of texts evidence a resurgence in the popularity of the living world motif, which appeared only infrequently throughout the 1950s. These proto-Gaian stories were part of the confluence of ideas that fed into the popular environmental movement and which informed sf's ecological discourse. The 1960s–1970s environmental movement also provided a context for understanding the Gaia hypothesis and for developing its latent mysticism. Stories belonging to this group include Richard McKenna's 'The Night of Hoggy Darn' (1964 [1958]) and 'Hunter, Come Home' (2001 [1963]), Le Guin's 'Vaster Than Empires and More Slow' (1982 [1971]) and *The Word for World Is Forest* (1976 [1972]), and James White's 'Meatball' (1971b [1969]) and 'Major Operation' (1971a). Lem's *Solaris* (2003 [1961]) was written in part as a response to American sf; for this reason and because it features a living world, this chapter explores its significance in the context of its 1970 English translation. Ernest Callenbach's *Ecotopia* (1978 [1975]), discussed at the end of this chapter, foregrounds the widespread impact of ecology during this period. It is with the proto-Gaian texts that this discussion begins.

1960s–1970s Proto-Gaian Living Worlds

McKenna's 'The Night of Hoggy Darn' and 'Hunter, Come Home' were both published before Lovelock formulated the Gaia hypothesis. They illustrate how certain sf themes shaped the motif of a living world that began to be considered in ecological terms as 'whole' and 'connected.' In 'The Night of Hoggy Darn,' a genocidal war between humankind and an indigenous dinosaur-like species of alien known as 'stompers' forms the core of the story. In 'Hunter, Come Home,' Ames the ecologist struggles against the aggressive Mordinmen to prevent them from unleashing a virus that could destroy an unnamed planet's ecosystem. 'Hunter, Come Home' rewrites the narrative of 'The Night of Hoggy Darn' in ways that reflect a shift in sf towards a transformative environmentalism that draws from countercultural politics.

Flinter Cole, the protagonist of 'The Night of Hoggy Darn,' explains that he is 'an ecologist – that means I deal with everything alive, and the way it all works in with climate and geography. I can use *any* kind of data' (McKenna, 1964, 9). Recognition of the interconnectedness of phenomena within an ecological system is part of Cole's holist approach to understanding the planet and the colonists. He borrows from a tradition of energy economics when linking this outlook to the theme of energy flowing through an ecosystem, explaining that he must 'make energy flow charts' documenting the interactions between life on the planet New Cornwall (9). The text explores the dimensions of a mythic-poetic response to its central conflict, what Cole describes as 'the greed-murder of species' because he believes it to be motivated by the colonial trade in stomper eggs, a delicacy that resonates with the historical exploitation of now extinct species on Earth. Cole learns that the stompers have been feeding on wild and captured humans; the logic of energy flows advanced by the text assigns to their eggs the status of 'human flesh at one remove' (42).

Two human groups inhabit New Cornwall: the 'subnormals who are so powerfully drawn to run back to the forests' and those colonists who retain a comparatively sophisticated technological culture and maintain contact with the wider interplanetary civilisation (McKenna, 1964, 41). The subnormals have developed their own spiritual explanation for the conflict between humans and stompers. Garth Bidgrass, head of Bidgrass Company, describes the subnormals' account of the conflict as 'a strange mixture of poetry and prophecy' and mythologises the colony's struggle with the stompers by casting it as a conflict between the archetypes Grandfather Man and Grandfather Stomper (41). Cole tells Bidgrass to '[t]hink of a species as one great animal that never dies, of which each individual is only a part' (41). Bidgrass connects Cole's ecologism to myth when he explains that 'your notion of the greater animal, critical biomass [...] [w]e speak of Grandfather Stomper and we are trying to kill him. He is trying to enslave Grandfather Man' (31). The concept of critical biomass possesses Gaian overtones and refers to the total mass of a particular species; if it falls below a critical threshold the species is unable to maintain a stable population and becomes extinct. Cole is caught in a narrative of prophecy, one that foretells how 'the new Grandfather Man will come naked out of the forest with his beautiful wife and armed with a thigh bone, that will lead us in the even greater task of reclamation that comes after' the death of Grandfather Stomper (58).

The links between ideas of ecological connectedness, proto-Gaian images of biomass and the foregrounding of archetypal figures illustrate the sf weaving of myth and science. 'The Night of Hoggy Darn' uses prophecy to transform Cole's experience of time into a narrative that relates, in mythic terms, a new sense of collective purpose and habitation. The mutual deaths of Grandfather Man and Grandfather Stomper in combat allow Cole the space to establish a new set of relations outside of the cycle of conflict on New Cornwall. As this space comes at the price of the genocide of the stompers, 'The Night of Hoggy Darn' accepts the necessity of conflict and human conquest, albeit in ways that are problematic. Bidgrass, for instance, admits that the early colonists 'killed off the great, stupid herds of darv cattle on which the stompers fed. The stompers that survived became wary and hostile' (McKenna, 1964, 38). The colonists are clearly the aggressors, which may partly account for why McKenna chose to rewrite this story as 'Hunter, Come Home.' In this narrative, he further develops the chronotope of the forest as a space of interconnection and transformation, demonstrating the development of a motif that is central to the Gaia theme.

In 'The Night of Hoggy Darn,' Cole becomes trapped in the forest and begins a 'fantastic journey [that] wound over great gnarled roots and buttresses fusing and intermingling until it seemed that the root-complex was one unthinkably vast organism' (McKenna, 1964, 50). McKenna makes this theme central to the narrative of 'Hunter, Come Home.' In this story two groups attempt to colonise a nameless planet, the Mordinmen's aggressive hunting society and the scientific Belconti. The planet is populated by 'phytos,' which the Belconti Midori explains to Craig, the Mordinmen protagonist, are 'mixed plant and animal. Life never split itself apart on this planet' (McKenna, 2001, 77). The two groups of colonists attempt but fail to displace the phyto ecology, which encourages the Mordinmen to deploy the Thanasis virus in an attempt to sterilise the planet. The Belconti Sidis explains that '[w]ith translocation, *Thanasis* can redesign its own free-systems in the field [...]. It could come up with something impossible to immunize, something no control virus we know how to make could handle. Then it would kill us and rule the planet itself' (73). Conflict occurs between a planet's whole ecology and human biological technologies, against which a smaller-scale struggle is played out between the constraining mores of Mordinmen society and Craig and Midori's developing romance.

Sidis notes that 'the phyto stems are all rooted together underground like one huge plant,' while Midori explains that '[the phytos] form a kind of biochemical intelligence, almost a mind, and it's learning faster than we are' (McKenna, 2001, 72, 77). Midori connects this visionary insight into the nature of the phytos to the theme of an alien consciousness emerging from the ecological system's interaction with an other: 'just think of the agony and the changings, through all the long years men have been trying to kill this planet. What if something ... somehow ...

suddenly *understands?*' (87). Midori connects love of people to a love of nature in one scene where, as Craig pilots a flyer over the forest, they witness a group of migratory phytos which, in Midori's words, 'stain the air with beauty! [...] It knows I love it' (86). Midori's affective response to nature's beauty in turn awakens Craig to a new appreciation of the phyto ecology.

Human connectedness to nature is literalised at the conclusion of this story through the concept of 'resorption.' Craig and Midori are exiled from the colonial settlement and, in a variant of reincarnation, they become infected with Thanasis only to have the phytos reconstitute their bodies and their consciousness molecule by molecule. Earlier, Midori tells Craig that 'this planet has never known death or decay. Everything is resorbed and reconstituted. We try to kill it and it suffers but its – yes, its *mind* – can't form the idea of death. There's no way to *think* death biochemically' (McKenna, 2001, 78). The energy flows that appear in 'The Night of Hoggy Darn' are thus refigured as Midori and Craig become continuous with alien nature in a version of the indistinguishability model of deep ecology, where the boundary between humankind and the rest of nature is erased. This continuity is joined to the theme of wholeness, generated through connections, and love: '[t]his life never split apart, Roy. In wholeness there is nothing but love' (96). Their rebirth is depicted as a state of Edenic innocence: '[s]he was smiling radiantly. They were both naked. He was not excited and not ashamed' (96). Midori raises the Romantic theme of the human observer's role in actualising the universe – as in William Blake's *The Marriage of Heaven and Hell* (1975 [1790]) – which makes humankind integral to a nature that can only conceive of itself through others' self-conscious reflection. She explains to Craig that

> [l]ike each littlest phyto, we are thoughts in that strange mind. I think we focus its awareness, somehow, serve it as a symbol system, a form-giver. [...] We are its thoughts that also think themselves, the first it has ever had [...] [i]t is a great and holy mystery, Roy. Only through us can it know its own beauty and wonder. It loves us and needs us. (McKenna, 2001, 97)

Midori privileges beauty and wonder as the two aesthetic responses towards nature most appropriate to human interaction with the phyto ecosystem. Burke and Kant support the link between beauty and love in their discussion of the beautiful, but the connection between wonder and need is more oblique. This network of linked impulses constitutes 'a great and holy mystery,' a state that resonates with Stan Godlovitch's emphasis on mystery as humankind's most prominent reaction to nature's otherness (1997). Echoing the theme of the inconceivable in Stapledon's *Star Maker,* Midori views the ecosystem as evolving towards the use of biotic life as a language to express its own otherness. Midori landscapes these expressions as thoughts, recalling nineteenth-century views of personified nature as a divine mind. Like Murray Leinster's living world in 'The Lonely Planet' (1949), the phyto ecosystem is dependent on interaction with otherness in the form of environed human consciousness to allow it to develop an awareness of its own processes. Humanity provides new symbols in the form of language, thought and emotion with which the ecosystem can continue to shape and express itself.

James White's Sector General stories 'Meatball' and 'Major Operation' hark back to the 1950s space opera and develop terraforming and proto-Gaian themes separately from the network of tropes established by McKenna and Le Guin. Their significance lies not in their contribution to the tropological complex for which the image of the living forest stands as an emblem, but in an alternative treatment of the terraforming and proto-Gaian intersection during this period. Nevertheless, connections to the chronotope of the forest are raised through description of life forms like 'great tracks of living "land" covered with the tiny, long-rooted plants which might or might not serve as the strata beast's eyes' (White, 1971a, 124).

Sector General Hospital is a multi-environment spaceship specialising in alien medicine; 'Meatball' relates an encounter with an alien species on the planet Drambo who appear to act analogously to leucocytes. They maintain the health of strata beasts, vast creatures the size of continents that have been under attack by indigenous aliens recently in possession of atomic power. In line with Sector General Hospital's mission to provide medical care to all sentient life, Dr Conway proposes a vast project aimed at preserving the life of the strata beasts. 'Major Operation' follows from the events of 'Meatball' after a surgical operation has been devised that utilises the medical benefits of the alien leucocytes discovered in the first story. In 'Major Operation,' Conway and the interspecies medical team discover that one of the strata beasts is intelligent.

The surgical procedure envisaged for the strata beast takes the form of an advanced engineering project: '[t]o a hypothetical observer ignorant of the true scope of their problem this medical treatment could have been mistaken for a very widespread mining operation, agriculture on an even larger scale and mass kidnapping' (White, 1971a, 152). The medical treatment of the land resonates with Lovelock's insistence on a science of geophysiology, with scientists operating as physicians aiding the planet to maintain a state of homeostasis. The choice of mining and agriculture as comparable activities to this surgical engineering programme forms a motivic series that has shaped the significance and meaning of terraforming throughout the tradition. Conway remarks that 'surgery on this scale will mean that the operation will be military rather than surgical,' which raises issues regarding the relationship between military and medical institutions, along with issues of scale and the capacity for human institutions to cope with projects of great magnitude (White, 1971a, 118). Use of military resources raises several ethical dilemmas regarding the motive for the project, while reference to 'mass kidnapping' highlights a questionable relationship between humankind and the various aliens on Drambo.

The altruistic provision of medical assistance, along with the opportunity to make contact with and perhaps learn from and teach other intelligent aliens, are initial justifications, but the project accelerates when it is discovered that there is an unknown intelligence able to manipulate the shape and function of 'tools' with its thoughts. These thought-malleable tools represent a resource of almost limitless application. It is soon discovered that the strata beast undergoing surgery mobilises them in self-defence, and only Conway's chance encounter and communication with the alien prevents further deaths amongst the humans. As 'Major Operation' ends, Conway struggles with guilt over the project's casualties: '[a] rather supercilious cultural contact specialist had tried to make it very simple for him by saying that difference, whether it was cultural, physiological or technological, was immensely valuable. They would learn much from the strata creature and the rollers while they were teaching them. Conway, with some difficulty, accepted that' (White, 1971a, 182). Conway's guilt raises an issue that substrates the use of the terraforming motif for both politico-cultural and environmental philosophical speculation. These Sector General stories explore the values and limits of peaceful contact with alien others, while the terraforming motif is used to call into question the status of the values that accompany humankind's intervention in alien affairs. The suggestion of resource exploitation and the potential military rather than medical application of these tools makes peaceful human–alien relationships ambiguous.

Lem's *Solaris* offers a critique of the anthropomorphism latent in scientific inquiry and the colonial conquest of space, themes central to understanding the enviro-ethical engagement that the imaginative spaces of terraforming and living world stories offer. *Solaris* reprises the motif of the alien planet covered by an ocean-like organism seen in Murray Leinster's 'The Lonely Planet.' Unlike Alyx, it is truly alien; the nature of its consciousness may never be definitively resolved. In contrast to Earth, the planet 'had reached in a single bound the stage of "homeostatic ocean"' and 'was capable of exerting an active influence on the planet's orbital path' (Lem, 2003, 118–19). There is a tantalising echo between Lem's sublime and grotesque depiction of Kelvin's reaction to the eponymous planet and the pan-psychic narrator's reaction to the sublime figure of the Star Maker, while the problem of analogising nature's otherness and the planet's mystery recalls Stapledon's portrayal of the Star Maker's radical otherness.

Although Lem praises Stapledon's *Star Maker* as 'a completely solitary creation' that 'defines the boundaries of the SF imagination,' he nevertheless finds fault with the 'double psychic nature of its ontology' (Lem, 1987, 7, 2). This flaw rests on the failure to resolve a contradiction between an image of the universe that exists teleomatically and possesses intrinsic worth and one that is transcendental and sanctified by a higher existence. Lem explains that deism leads to a 'behaviouristic teleology' in which 'we can only reconstruct the axiomatics of choices relevant to the highest values from the Star Maker's behaviour' (3). The ambivalent figure of the Star Maker symbolises this contradiction between a material and transcendental view of the cosmos.

Despite decades of Solarist studies that proliferate explanatory theories and scientific literature, the psychologist Kelvin reflects that humanity is no closer to understanding the planet: 'our scholarship, all the information accumulated in the libraries, amounted to a useless jumble of words' (Lem, 2003, 23). This extended confrontation with the alien, as with Alyx in 'The Lonely Planet,' recapitulates sf's autologic theme. *Solaris* exposes the personification also undeniably latent in Lovelock's naming of the Gaia hypothesis as an attendant problem in human attempts to describe nature. Lem reacts against the American pulp tradition of contact as equal exchange by 'creating a hall of mirrors with no windows from which to observe some privileged non-corresponding structure of things' (Csicsery-Ronay, 1985, 12). Kelvin points out that '[c]ontact means the exchange of specific knowledge, ideas, or at least of findings, definite facts. But what if no exchange is possible? If an elephant is not a giant microbe, the ocean is not a giant brain' (Lem, 2003, 152). *Solaris* does not offer an insight into a proto-Gaian consciousness so much as it destabilises human attempts to acquire knowledge of the cosmos by withholding the vantage of an external perspective outside of the structures of scientific inquiry by which to establish certainty. It reopens a space for the entrance of nature's otherness into the text by recourse to an aesthetic that Godlovitch calls a 'sense of mystery' in response to nature's elusiveness (1994, 26).

The moment where exchange between humanity and Solaris seems closest occurs through a dream shared by Kelvin and the planet after it is bombarded with X-rays modelled after his mind structure. Kelvin dreams that he has become a 'prisoner of an alien matter,' or rather that his identity has become indistinguishable from the ocean planet: 'I have no body, I *am* that alien matter' (Lem, 2003, 187). This dream is located 'in the heart of vastness,' in a space that denies orientation and volition (186). Recalling the duality of mind in *The Purple Cloud* and 'The Earth-Brain,' identification with the planet offers an experience with otherness that exceeds a narrowly human perspective, leaving him feeling that he 'has just left a state of true perception' (187). As the dream progresses, a splitting occurs in the indistinguishable consciousness and Kelvin's ego is simultaneously created and recreated, thus anticipating McKenna's portrayal of a recreated ego in 'Hunter, Come Home,' published two years after the original Polish edition of *Solaris*. This split ego in turn creates another: '[w]e discover one another mutually, beyond any effort of will, in an absorbed silence. I have become alive again, and I feel as if there is no limitation on my powers. This creature – a woman? – stays near me, and we are motionless' (Lem, 2003, 187). Csicsery-Ronay suggests that this scene represents the possibility of some form of contact, and proposes that the figure is Rheya, created in a moment of co-operation between Kelvin and the planet (1985, 10–11). Interpretation is foreclosed by the structure of the novel, suggesting the possibility that these dreams and projections simply represent an apparently inescapable act of landscaping when confronted with the ineffable. Like Stapledon's narrator, Kelvin attempts to describe his dream 'in so far as my vocabulary permits, given that I can convey only fragmentary glimpses almost entirely denuded of an incommunicable horror' (Lem, 2003, 186). Vastness, terrible silence and the horror associated with the dream point to an experience both sublime and grotesque and couple creation to the figure of the living world.

Kelvin and the scientist Snow discuss a belief in an 'imperfect god,' a 'sick god, whose ambitions exceed his powers and who does not realize it at first' (Lem, 2003, 206). Snow describes this idea as 'an evolving god' and suggests that this notion could apply to humanity, but Kelvin argues that it is 'an anchorite, a hermit of the cosmos, not a god,' and speculates that '[p]erhaps he has already been born somewhere, in some corner of the galaxy, and soon he will have some childish enthusiasm that will set him putting out one star and lighting another' (207). These explanations echo the narrator's attempt in *Star Maker* to comprehend the nature of the cosmos: they speak to the failure of the scientific team in their attempts to understand the nature of Solaris, part of a project of scientific enquiry lasting over a century that yields little in the way of understanding or possibilities for contact with the alien. Snow desperately speculates that 'perhaps it wants to please us but doesn't quite know how to set about the job,' recalling Leinster's and Bradbury's depiction of proto-Gaian planets whose agency is directed towards the fulfilment of human desire (191–92).

Human responses to cosmological nature, according to Snow, typically disrespect its alien otherness. Rafail Nudelman suggests that in Lem's sf, 'man is a tool of the self-cognizing cosmos to the same extent that wandering in the cosmos' labyrinth is a means of finding out about himself' (2000, 183). The emergence of consciousness in living world stories such as Leinster's and McKenna's and *Solaris* explores ways in which humankind relates to the otherness of the cosmos, and how the cosmos itself interacts with the human as its other. Snow explains the metanarrative of space colonisation as a strategy of avoidance of the other on our doorsteps rather than the desire for an encounter with not only alien nature, but the familiar otherness of nature on Earth: '[w]e don't want to conquer the cosmos, we simply want to extend the boundaries of Earth to the frontiers of the cosmos. [...] We are only seeking Man. We have no need of other worlds. We need mirrors. A single world, our own, suffices us; but we can't accept it for what it is' (Lem, 2003, 75). Failure to accept Earth by extending the boundaries of human identity to crowd otherness out is a psychological strategy of colonial imposition. Solaris, however, continually defies not only understanding, but analysis. Otherness reveals the limits of humankind's explanatory devices; the scientific project is imagined not as a march towards understanding but as a slower 'stumbling, one- or two-step progression from our rude, prehistoric, anthropomorphic understanding of the universe around us' (Lem, 2003, 178). The planet remains incomprehensible; 'an unbreakable silence' is Solaris's continued answer to any attempt to make contact, thus fostering the many theories that explain it according to 'the widely held notion [...] that the "thinking ocean" of Solaris was a gigantic brain [...] a sort of "cosmic yogi," a sage, a symbol of omniscience' (24). Lem argues that this silence in response to humankind's attempt to apprehend the nature of the cosmos is also the ultimate result of Stapledon's treatment of the figure of the Star Maker: he 'directs the traditional drama [of cosmogonic enquiry] by manipulating not-entirely-traditional figures and symbols' (1987, 7). Encounters with non-human others instigate a sense of mystery towards nature that decentres humankind's geocentric perspective.

Alluding in its title to Andrew Marvell's 'To His Coy Mistress' (2009 [1681]), Le Guin's 'Vaster Than Empires and More Slow' features an interconnected proto-Gaian forest that recalls McKenna's forest landscapes and Lem's depiction of radical alien otherness and human interaction with that nature. A small group of colonists land on 'a pure phytosphere' that they designate the planet 4470 (Le Guin, 1982, 33). Mannon argues that the root nodes of the forest they have been studying there 'are, indubitably, connections. Connections among the trees. [...] they are all interconnected, both by the root-node linkage and by your green epiphytes in the branches. A linkage of incredible complexity and physical extent' (50). Harfex argues that the forest is 'merely a network of processes. [...] they must all be capable of transmitting electro-chemical impulses. [...] Even the pollen is part of this linkage [...] a sort of windborne sentience, connecting overseas [...] the biosphere of a planet should be one network of communications, sensitive, irrational, immoral, isolated' (Le Guin, 1982, 54). 'Vaster Than Empires and More Slow' is an important intervention into the proto-Gaian narrative for its reconceptualisation of the living world in terms of ecological principles. Planet 4470's isolation, its closed global-scale system and the emphasis on communication characterise the planet as a cybernetic system. Osden has Render's syndrome, making him empathic but unable to control his ability to reflect the distorted and strengthened emotions of others. This dynamic recalls Lem's portrayal of the living planet Solaris and parallels the forest's response to the colonising expedition; as an isolated entity, it has never had the opportunity to respond to otherness.

Osden remarks that the forest is '[o]ne big green thought' (Le Guin, 1982, 54). The other colonists note 'a hypnotic quality in the colors and spacing of the stems and branches' and question whether 'the forest ambiance has a rather troubling and possibly hallucinogenic effect of the perception' (38). Mannon points out that he 'feel[s] a strong anxiety with a specific spatial orientation [...] [for] which the archetypal connotations of the word 'forest' provide an inevitable metaphor' (47). The forest is a space that evokes a primordial terror which – like the hate and fear that Osden exacerbates – is a reflection of the affective response of the colonists to the alien forest. Osden's answer to the positive feedback of this cycle is to sacrifice himself by submitting to the fear that the forest amplifies. Thus he hopes to establish communication with the forest and transmit to it another signal based on a structure that, from the human perspective, is implicit at the scale of cosmological phenomenon: '[a] single human brain can perceive pattern on the scale of stars and galaxies [...] and interpret it as Love' (55). Individuals are thus able to translate these vast spaces into terms that can be understood at smaller scales of experience.

Le Guin uses the proto-Gaian forest as a motif to examine processes of othering. The narrator explains that Osden 'had given up his self to the alien, an unreserved surrender, that left no place for evil. He had learned the love of the Other, and thereby had been given his whole self' (Le Guin, 1982, 58–59). Val Plumwood discusses one axis of the logical process of dualism, radical exclusion, which works to 'maximise distance or separation between the dualised spheres and to prevent their being seen as continuous or contiguous' (1993, 49). Accordingly, Osden's 'whole self' is granted him only when he overcomes the fear and hatred of the other and accepts what is alien. By bridging the distance between dualised concepts, love becomes a strategy for overcoming radical exclusion. In this light, the opposition between humankind and stompers in 'The Night of Hoggy Darn' and the humans and phytos in 'Hunter, Come Home' can be seen as expressions and negotiations of this logic of dualism. Crucially, the dualised terms are different in as many ways as is possible: '[d]ualistic distinction aims to maximise the number, scope, or significance of distinguishing characteristics' (Plumwood, 1993, 50). The 1960s popular ecological notion of love as it has been incorporated and rendered in Gaian sf stories can be seen as a metaphorical attempt to effect a synthesis between hyperseparated terms of a dualism in order to overcome the self-destructive tendencies underlying humankind's approach to nature and other cultures.

Terragouging: Time and the Forest

Myth is central to connecting the cultural attitudes and behaviour of groups and the treatment of nature in Le Guin's *The Word for World Is Forest*, a Vietnam-era story of a colonising force of Terrans who subjugate the indigenous population of Athshe and deforest the planet to lessen Earth's deficit of this scarce resource. The theme of energy flowing through an ecosystem draws from a tradition of New Age spiritualism influenced in part by a romantic conception of the myth systems of tribal cultures. Le Guin's oeuvre draws from Native American culture and myth, an interest that can be traced back to her father, the influential anthropologist Alfred L. Kroeber, and her mother, Theodora Kroeber, whose biography, *Ishi in Two Worlds* (1973 [1961]), helped shape conceptions of Native American mythology and popularised the idea of their essential connection to nature. Her brother, Karl Kroeber, was an academic, whose scholarship on the English Romantics and on American Indian literature likewise influenced her fiction.

Captain Davidson believes that 'primitive races always have to give way to civilised ones. Or be assimilated. But we sure as hell can't assimilate a lot of green monkeys' (Le Guin, 1976, 12). He associates nature, cultures lacking techno-scientific infrastructures and the animal in a network of dualities that for him justifies the exploitation of the forest and the enslavement of the Athsheans. This attitude is shared by many of the other colonists: Colonel Dongh affirms that 'they're not human beings in my frame of reference' (12). The Athsheans understand the threat posed by the colonists against a religio-spiritual background centred on their practice of 'dreaming,' a scientifically measurable phenomenon that Captain Raj Lyubov, an anthropologist sympathetic to their plight, manages to detect while working with the enslaved Athshean Selver. Once freed, he explains to his people that '[the colonists] have left their roots behind them, perhaps, in this other forest from which they came, this forest with no trees,' and concludes that '[n]o one can say certainly whether they're men or not men, whether they're sane or insane, but that does not matter. They must be made to leave the forest, because they are dangerous' (44–45). Insanity as characteristic of the colonists' treatment of the environment is contrasted with the indigenes' saner, spiritual relationship and works to critique the instrumental rationalism underlying the colonial logic of othering.

Dreaming is deeply involved in the Athsheans' perception of time, which in their world is intimately connected to their sense of place in the forest. There is no sharp boundary between dream time and the world's time in Athshean thinking, unlike the colonists' demarcation between these spheres along an unreal/real axis. The space of the forest itself is oneiric and challenges the instrumental version of progress that informs the colonists' terragouging of the planet. As Lyubov learns more about this aspect of their culture, he comes to believe that this sense of time is central to any basic understanding of the Athshean worldview: '[i]t was Selver who had made him understand, at last, the Athshean significance of the word "dream," which was also the word for "root," and so hand him the key of the kingdom of the forest people' (Le Guin, 1976, 100). The words 'dream' and 'root' are connected semantically, and the words 'root' and 'forest' metonymically, and because, as Lyubov claims, '[t]he Athshean word for *world* is also the word for *forest*,' their cultural life is revealed to be intimately connected to their environment (72).

The clash between Selver and Davidson represents two sides of a dualism that for Selver is self-conscious. During their final confrontation, Selver tells Davidson that 'we're both gods, you and I. You're an insane one, and I'm not sure whether I'm sane or not. But we are gods' (Le Guin, 1976, 160). This view mirrors the self-conscious use of mythic archetypes in 'The Night of Hoggy Darn' and invests this clash between cultural representatives with a sense of urgency that accompanies the development of a new cognitive relation to the world. Time is compressed as the Athsheans' experience as members of a subject race is focused upon Selver. The Athshean word *sha'ab* for god also means 'translator'; through the act of translating dreams into reality the *sha'ab* brings a gift to his people and thus becomes a god. Lyubov ponders whether Selver '[m]ight [...] then be one who could translate into waking life the central experience of vision: one serving as a link between the two realities, considered by the Athsheans as equal, the dream-time and the world-time, whose connections, though vital, are obscure' (106). Through active dreaming, the submerged desires and fears of the people as a whole can be confronted and, given sufficient need, brought into the time of the real world. An old Athshean dreamer claims that '[w]e may have dreamed of Selver these last few years, but we shall no longer; he has left the dream-time' (34–35). The Athsheans have called forth a god in response to the thirty years of colonisation and exploitation they have endured, one that will translate their new cultural experience into terms they are capable of responding to, primarily by effecting a change in Athshean culture.

Davidson, on the other hand, is a god who is unaware of his power to make manifest his dreams. Selver comments of the colonists in general that '[i]f they are men they are evil men, having denied their own gods, afraid to see their own faces in the dark' (Le Guin, 1976, 45). These gods can be understood as cultural tendencies that have been repressed and disavowed, thus resulting in the monstrous emergence of practices that lead to the exploitative treatment of nature and the indigenes. The officers' reports on their impact on the indigenous ecosystem and culture during a conference held between the military authorities and two alien emissaries from the newly formed Council of the League of Worlds demonstrate the extent of this denial. Colonel Dongh insists that '[w]e do not employ slaves, sir. Some of the natives serve a useful role in our community. The Voluntary Autochthonous Labor Corps is a part of all but the temporary camps here' (63). Lyubov is the only officer to provide an oppositional account of their actions: '[w]e have killed, raped, dispersed, and enslaved the native humans, destroyed their communities, and cut down their forests. It wouldn't be surprising if they'd decided that we are not human' (62). In Le Guin's Hainish universe, all conscious humanoid life on all of the known planets descended from an original Hainish stock. This fact blurs the boundary between the human and non-human sides of the dualism and undermines the related alien/non-alien dualism that informs the colonists' treatment of the Athsheans. It is significant that Colonel Dongh refuses to accept this idea as 'the historic fact,' illustrating another denial of shared characteristics between the Athsheans and the colonists and marking the former from the latter as a dualised other (64).

Selver and Davidson are manifestations of their cultures' orientation towards nature and other cultures. This relationship echoes McKenna's earlier use of the mythic figures Grandfather Man and Grandfather Stomper, both of whom are microcosms and representatives of their society and species. Davidson's and Selver's roles are instrumental in defining the shape of their cultures' future engagement with the world and each other. Selver explains to Davidson that '[y]ou gave me a gift, the killing of one's kind, murder. Now, as well as I can, I give you my people's gift, which is not killing. I think we each find each other's gift heavy to carry' (Le Guin, 1976, 160). Contact with a violent culture has led the Athsheans to adapt by appropriating violence, and once incorporated into their cultural consciousness it cannot be shed. Recalling Osden's relationship to the forest in 'Vaster Than Empires and More Slow,' the arrival of the colonists disrupts the idyll of the forest and instigates a feedback cycle between the two cultures, forcing the Athsheans to respond, not with love, as Osden does in 'Vaster Than Empires and More Slow,' but with violence. This results in the Athsheans' loss of their greatest cultural triumph, which the Hainish emissary Lepannon identifies as '[a] human society with an effective war-barrier!' (61). This 'war-barrier' is overcome at the conclusion of the story as the violence towards and reprisals by the Athsheans establish a positive feedback loop between the two cultures. The Athsheans are 'driven by the evil dream [of killing] and only Selver could teach them how to master it' (113). Selver, however, is doubtful that they can return to a time predating the internalisation of this violence: the clash of cultures has inaugurated a sense of time that is linked to the beginning of a new phase in the Athshean worldview.

Terraforming and proto-Gaian stories combine sf's religio-spiritual language with a scientific perspective to develop a series of ecological narratives that explore alternative conceptions of nature. Important for this ecological branch of sf is the development of proto-Gaian motifs and the representation of low-technology cultures living closely with nature. McKenna and Le Guin developed proto-Gaian themes of a mythic-poetic character that represents one way in which sf explores confrontations with a living landscape. Another is the radical exclusion of indigenous populations from the values associated with a colonial culture. McKenna depicts a mythic struggle between the indigenous and recognisably non-human stompers and the colonists who have lived for generations on the alien planet. In contrast to the pattern of conflict that dominates space opera, White explores the problems that arise as a consequence of peaceful yet potentially fraught contact with alien others. The intersection of proto-Gaian themes and the trope of indigenous alien life as the object of an affective response creates a space where these motifs are flexibly deployed to explore human relationships to the non-human world in a variety of contexts. Underlying many of these stories is a spiritual orientation to landscaping the non-human. Indigenous alien societies or living worlds are often exemplars of a pseudo-pantheistic religion through which a connection to nature's otherness has been made.

Alongside these indigenous populations and colonial cultures are characters living on the fringes of their social groups. As outsiders, they come into contact with the indigenous as the site of projected values that are seen as absent from Earth and its societies. The space created by the clash between cultural systems brings these individuals to a heightened awareness of the beauty and sublimity of the non-human world, and their affective responses lead them to quasi-spiritual modes of perceiving and characterising nature. Sf thus develops a language that puts into conflict religio-spiritual and rationalised exploitative modes of understanding the world to explore a non-human nature that exceeds its status as a resource to be used or a space to be colonised. Scientific modes of perceiving the world, connected to but distinct from rationalised exploitation, work to support these spiritual conceptions of nature.

Terraforming in the 1960s–1970s

Herbert's *Dune* sequence, Heinlein's *The Moon Is a Harsh Mistress* and Le Guin's *The Dispossessed* do not depict living worlds, but rather use terraforming to explore the political landscapes developed by earlier terraforming stories, reconsidering them in the light of the wider cultural shifts of the 1960s–1970s. As with terraforming in the 1950s, extreme environments encourage the formation of tightly knit communities where co-operation is central to survival. There is a struggle between this sense of community and the perception and use of planets as prisons or sites of oppressive government control, as Poul Anderson portrays in his nightmare vision of isolationist dystopias.

The texts considered here rely on a political contrast that opens a space for contesting voices to emerge. In *Dune*, the society in question is a vast interplanetary empire constituted of ducal fiefdoms; Heinlein's *The Moon Is a Harsh Mistress* contrasts Earth as a colonising centre with the terraformed Moon. In Le Guin's *The Dispossessed*, the political opposition between Urras and its moon is complicated by the historical concession of the moon (Anarres) to the rebel Odonians to prevent further political destabilisation at the centre. Recalling *Dune* and the interplanetary societies in the space opera of the 1950s, Le Guin situates her exploration in her Hainish universe, thus offering further contrasts with a wider group of worlds. These works, like those of the 1950s, borrow from the utopian tour, whereby a visitor to utopia is guided by an inhabitant in surveying the alternative society, thus allowing that society to be assessed from an outsider's perspective. While Heinlein's main character – Manuel Garcia O'Kelly – is already integrated into this society, the first-person address makes the target of the initiation the implied reader.

The political dimension of the theme of interconnectedness was anticipated by Wells's link between the physical landscape and human nature on the one hand and by Heinlein's interest in population bionomics in *Farmer in the Sky* on the other. The transformations that social governance brought to culture and the environment, often abstracted into ecological issues such as overpopulation and resource scarcity, continued to be popular throughout the 1960s–1970s. Herbert, Heinlein and Le Guin mobilise ecological principles in ways that connect scientific ideas to society, establishing an ecopolitical context that later terraforming stories would develop and dovetail with popularisations of Gaia. Heinlein's *Stranger in a Strange Land* (1961) became the unofficial bible of the counterculture, and his work, especially *The Moon Is a Harsh Mistress,* would greatly influence Brand and the pragmatic ecologism that cohered around the *Whole Earth Catalog.* The relationship between ecology and politics and the notion that human landscapes form a necessary part of ecological thinking are now widely accepted in environmental philosophy and were already prefigured in the work of Aldo Leopold, Pierre Teilhard de Chardin and other environmental philosophers and scientists. This cluster of texts illustrates a shift in the character of the political engagement of stories dealing with terraforming, pointing to a response on the writers' part to the political assumptions of earlier sf.

Terraforming and Ecopolitics in the *Dune* Sequence

Herbert attributes the initial inspiration for the *Dune* trilogy to an uncompleted magazine article, 'They Stopped the Moving Sands,' which focused on the US Department of Agriculture's project to cultivate poverty grass to bind sand dunes in Florence, Oregon. Martyn Fogg writes in his preface to *Terraforming* that the Nebraska Sand Hills, having undergone a similar treatment, provides a *'marvelous metaphor* [...] *for terraformers,'* and he speculates, '[w]*hat if we could engineer the sand seas of Mars'* (1995, xii). These connections between geoengineering and terraforming, between transformations of Earth and other planets, are typical of the ecological concerns of this type of narrative. Herbert writes, 'I could begin to see the shape of a global problem, no part of it separated from any other' and that '[a] new field of study rises out of this like a spirit rising from a witch's cauldron: the psychology of planetary societies' (2007). As one of the most influential examples of ecological sf, the *Dune* trilogy's treatment of nature and its holism – which factors human worlds into ecological systems – have informed sf discourse and influenced popular culture, thus helping to shape sf's future ecological vision.

Dune begins with the ducal family Atreides, who take possession of the Arrakeen feudal house, the political centre of the planet Arrakis and its spice-mining operations. It depicts a politico-economic struggle over control of the planet, fought between the interplanetary Empire and the indigenous Fremen, led by Paul Atreides. Throughout the *Dune* trilogy the spreading growth of plant life on the desert planet stands as an emblem for Fremen dreams of freedom, abundance and vitality. As Arrakeen's influence grows after Paul becomes Emperor at the end of *Dune*, this dream is threatened by the destruction of an older form of spiritual cohesion and reciprocity with the planet, symbolised by the threatened disappearance of the deep desert, the sandworms, and the supremely valued spice. As *Dune* concludes, Paul imposes his vision of a future Arrakis onto the Emperor he has overthrown: '"The Fremen have the word of Muad'Dib," Paul said. "There will be flowing water here open to the sky and green oases rich with good things. But we have the spice to think of, too. Thus, there will always be desert on Arrakis ... and fierce winds, and trials to toughen a man["]' (Herbert, 1965, 462). This image encapsulates the contrast between the wilderness of the present Arrakis and the promise of a pastoral synthesis between civilisation and nature. Paul's oath is a political act that focuses the inhabitants' efforts towards the shaping of a new future. The wilderness is valued not just as an economic asset but for its role in developing individuals that can be used as tools to enforce Paul's sovereignty. Those who survive do so because of their rigid discipline, their subordination to the welfare of the group and their individual strength of character, traits aligned with a heroic militarism imagined as an exclusively masculine domain. James Oberg cites *Dune* as exemplary of the examination of the role of people in terraforming (1981, 120), while the theme of individuals and groups who are used as tools continues to be an important influence on terraforming narratives such as Pamela Sargent's *Venus of Dreams* (1989a [1986]).

The planet's importance as the sole source of spice provides the incentive for this instrumental use of communities and the land. Liet Kynes is the Imperial Planetologist, a position that, under Harkonnen rule, amounted to the use of the *'native labour pool'* to conduct the terragouging of Arrakis (Herbert, 1965, 26). Liet recalls his father Pardot describing this political arrangement in economic and hierarchical terms: '"Arrakis is a one-crop planet," his father said. "One crop. It supports a ruling class that lives as ruling classes have lived in all times while, beneath them, a semi-human mass of semi-slaves exists on the leavings. It's the masses and the leavings that occupy our attention. These are far more valuable than has ever been suspected."' (Herbert, 1965, 262). Pardot's scheme for terraforming Arrakis seeks to tie the Fremen's emancipation to a long-term project of world construction that builds multiple levels of physical and cultural adaptation into a single transformative network. Pardot believes that, like the physical processes that have shaped Arrakis, '[o]ur timetable will achieve the stature of a natural phenomenon [...] A planet's life is a vast, tightly interwoven fabric' (Herbert, 1965, 263). The new community that Pardot seeks to develop is subordinated to the directive of the messianic hero Paul and, in *Children of Dune*, to his grandson Leto II. This is accompanied by a shift in the timescale envisioned for terraforming and emancipation, which falls from geologic to generational spans. Anticipating the acceleration that Paul's escape into the deep desert in *Dune* promises for the ecopolitical project, Pardot warns Liet that '[n]o more terrible disaster could befall your people than for them to fall into the hands of a Hero' (Herbert, 1965, 263). This, as Herbert remarks in 'Dune Genesis,' is the originary concept for the trilogy: 'the messianic convulsions that periodically overtake us' (2007). The Fremen 'masses' are subordinated to the agenda of an individual aristocrat. The terraforming project, originally conceived as the slow growth of a culture co-adapting with its environment, is commandeered for the purposes of an individual's drive for revenge and power. The theme of heroic individualism can be traced back to 'Born of the Sun' (1934) and, as Susan Stratton contends, is one of the enduring sf tropes that pose problems for ecocriticism (2001).

Ecology is central to the terraforming motif in *Dune* and provides a conceptual bridge between concern for the natural world and an examination of the groups who inhabit the planet. Plans for terraforming Arrakis are overseen by Liet, whose father Pardot, as Arrakis's planetologist before him, supplied the ecological vision for the long-term terraformation of the planet. Liet continues to hallucinate his father's early teaching on ecological principles when abandoned by the Harkonnens in the deep desert: '[w]e are generalists [...] You can't draw neat lines around planet-wide problems. Planetology is a cut-and-fit science' (Herbert, 1965, 260). The title 'planetologist' signals a shift away from the specificities of local ecosystems to a focus on the global. Pardot first identified the potential for ecological management offered by the nomadic Fremen, noting that '[t]o the working planetologist, his most important tool is human beings [...]. You must cultivate ecological literacy among the people. That's why I've created this entirely new form of ecological notation' (260). The importance placed on ecological literacy in the context of the directed modification of a planet links ecological education to projects for human emancipation. This vision of an endeavour that includes both natural and cultural worlds, and the possible transformations that can be effected by a global community, is expanded when Liet recalls his father's visionary ambition: '[w]e must do a thing on Arrakis never before attempted for an entire planet [...] We must use man as a constructive ecological force – inserting adapted terraform life: a plant here, an animal there, a man in that place – to transform the water cycle, to build a new kind of landscape' (261). This passage links a nomadic, globalised movement over the land to a conscious, directed transformation of social and cultural practices and recalls Wiener's reframing of human populations in cybernetic terms in *The Human Use of Human Beings*. Pardot's call for a new landscape oscillates between natural and cultural referents and leaves the realisation of this landscape ambiguous, demonstrating how such global transformations are dependent on the interrelation of nature and culture.

This focus on planetary ecology strongly anticipates Lovelock's interest in Gaia as a metaphor for considering Earth as a planetary system. When reflecting on how the sandworms' evolution and life cycle shape the planet, Pardot notes that 'the near-ideal nitrogen-oxygen-CO2 balance [is] being maintained here in the absence of large areas of plant cover' and claims that '[t]he Arrakeen environment built itself into the evolutionary pattern of native life forms,' both these statements echoing elements of Gaian processes and effects (Herbert, 1965, 261). Pardot's claims also place humankind in relation to a global environment where nature is managed for instrumental ends. For Pardot, the Fremen embody 'an ecological and geological force of almost unlimited potential' (467). Vegetative life can establish homeostatic regulation of the planet once a critical biomass is reached. The Fremen capitalise on this process to transform Arrakis by establishing a new set of parameters that they initially modulate until a new regime stabilises. These principles are applied to socio-political relationships. Liet, who constantly thinks of the Fremen using the possessive 'my,' Pardot, who cares nothing for the individual Fremen over the group and who is content to '[l]et them think anything they wish as long as they believe in us,' and Jessica, who speculates that the Fremen 'could be wielded like a sword to win back Paul's place for him,' all view the Fremen instrumentally (263, 304). Social engineering is given the status of an ecological principle: '[m]ovement across the landscape is a necessity for animal life,' Pardot explains. 'Nomad peoples follow the same necessity. Lines of movement adjust to physical needs for water, food, minerals. We must control this movement now, align it for our purposes' (261). The migratory lifestyle that informs the Fremen's culture is, in ecological terms, a flow of energy that can be harnessed to direct intervention into the environment.

This migratory lifestyle is at odds with Kynes's realisation that terraforming requires another kind of civilisation: '[a] thought spread across his mind – clear, distinct: *The real warmth of a planet is in its landscape, how we take part in that basic source of civilization – agriculture*' (Herbert, 1965, 259). Agriculture is linked to terraforming and necessitates the development of a rooted culture. Opposing this vision of an agricultural landscape is that of the migratory sandworms, who, like the Fremen, embody a vast force aligned with movement: Pardot explains that '[i]t was lines of movement that gave us the first clue to the relationship between worms and spice' (261). Two cultural forms of habitation symbolised by agriculture and the sandworms are thus opposed, and a synthesis negotiated. When Paul is adopted into a Fremen tribe he realises that he 'was surrounded by a way of life that could only be understood by postulating an ecology of ideas and values' (329–30). These cultural systems are metaphorical ecologies that enframe the physical and cultural parameters of a planet and its population.

Arrakis's global environment, the chronotope of the barren desert wilderness, offers advantages for ecopolitical reflection and the growth of an eco-cosmopolitanism that continues to be explored in other works of terraforming. Arrakis's harsh environment fosters a heightened awareness of the impact of the environment on the body and, at a larger scale, the constraints to the development of indigenous communities and civilisations. The scarcity of resources on this planet brings the economic basis of humankind's relationship to the environment into focus: '"[t]he historical system of mutual pillage and extortion stops here on Arrakis," his father said. "You cannot go on for ever stealing what you need without regard to those who come after. The physical qualities of a planet are written into its economic and political record. We have the record in front of us and our course is obvious."' (Herbert, 1965, 262). The wasteland chronotope is ideally suited to highlighting the implications of an economic system that operates by exploiting others for access to an ever-dwindling supply of resources. A sense of time geared towards responsibility to future generations is joined to this outlook, a new awareness that can be considered a form of eco-cosmopolitanism that struggles with the repercussions of the human impact on the landscape and the changes to how the landscape signifies for the Fremen. Arrakis's environment does not determine so much as provide the initial foundations for the economic and political systems that Paul establishes. The character of this ecological vision is undermined by the dominance of the main heroic narrative and its strong instrumental approach to nature, but it does remain a critical undercurrent in the sequence.

The Garden in *Dune*

Jessica discovers a private garden in the Arrakeen feudal house that gratuitously consumes enough water to support many people. This garden more than the lack of waterseals on doors and windows is 'a deliberate statement of power and wealth,' signified by the callous and calculated waste of water on a world that lacks this resource (Herbert, 1965, 72): '[w]ater everywhere in this room – on a planet where water was the most precious juice of life. Water being wasted so conspicuously that it shocked her [Jessica] to inner stillness' (Herbert, 1965, 72). The garden is no refuge but a symbolic affirmation of aristocratic power and wealth by the Harkonnens and, by extension, the Empire and its overarching control over Arrakis's future. This space has, however, been appropriated for other purposes. A note left by the Emperor's proxy, a fellow aristocrat and Bene Gesserit, allows Jessica to locate a hidden message warning of an assassination attempt on Paul, who appears with a 'hunter-seeker' before she can act on the warning and, at her direction, destroys the mechanical assassin by plunging it into a pool. The ensuing dialogue between the two constructs another layer of significance that refigures this garden space as a refuge from further threat. This is emphasised by the airlock sealing the garden from Arrakis's arid atmosphere. Its special climate provides a contrast to the desert landscape it excludes and shows how the trope of the airlock acts as both a threshold and a discontinuity between these spaces. The garden is represented as another world through a series of technologies that allow the climate to be modified, including the aforementioned airlocks, the filter glass that simulates a yellow sun and the 'clock-set servok,' a sprinkler that connects images of technology, water and the garden. Whereas danger is aligned with technology through the hunter-seeker's function as a counterforce and this technology is symbolically contrasted with the natural through its destruction by water, the garden is completely dependent upon technology for its existence (Herbert, 1965, 71–74).

This garden landscape is a politically contested space. In contrast to the initial power statement it originally signified under Harkonnen feudalism, the Atreides co-opt it to underscore their ideological difference and to legitimate their own rule: '"My Lord, the Duke, and I have other plans for our conservatory," Jessica said. She smiled at Leto. "We intend to keep it, certainly, but only to hold it in trust for the people of Arrakis. It is our dream that some day the climate of Arrakis may be changed sufficiently to grow such plants anywhere in the open."' (Herbert, 1965, 127). Kynes interprets this promise against the Fremen legend of '"the shortening of the way,"' the appearance of a mythical hero known as the Kwisatz Haderach, who would accelerate the terraforming of Arrakis. Jessica is prompted to wonder, *'Did our Missionaria Protectiva plant that legend here, too?'* (Herbert, 1965, 128). The garden becomes an ambivalent symbol for freedom from oppressive feudalism. If the garden came to be mirrored by the wider landscape, then the scarcity and control of water on Arrakis could no longer bolster the economic power exercised over the Fremen. However, in a text so dominated by the ceaseless plotting between feudal houses, the co-opting of the pastoral ideal operates as a counterforce that introduces civilisation – aligned with the Empire – into the ideal. The cynicism with which Jessica exploits Fremen legend and the politically charged statement that echoes Arthur C. Clarke's *The Sands of Mars* (1951), 'hold it in trust for the people,' dialogises the different discourses of the text. *Dune* draws further correspondences to real-world historical events via the spice industry's metaphorical connection to oil, clean air, water and other resource shortages that occur as a result of aggressively exploitative commerce.

The garden remains alien to Arrakis. When it is revealed that free water on the planet's surface threatens the sandworms' extinction, the pastoral ideal that the dream of a terraformed planet offers to the Fremen is gradually undermined. In *Children of Dune*, the Atreides retainer Gurney Halleck views the new landscape that Paul's vision brought to the desert and is able to assess it from the vantage offered by his status as an outsider: '[m]omentarily, he saw the garden through Fremen eyes: alien, menacing, dangerous in its waste of water [...] *Both of us are alien here*' (Herbert, 1976, 334–35). The speed of the planet's physical transformation creates a disjunct between Halleck and his environment, an estrangement that is shared by the garden; like the offworlder who views this landscape, the garden superimposed onto Arrakis is alien to its environment. This sense of alienation can be usefully compared to notions of deterritorialisation that Heise discusses in *Sense of Place and Sense of Planet* (2008). Deterritorialisation involves the emergence of new cultural forms and practices that are no longer rooted in place. In the *Dune* trilogy this deterritorialisation hinges on the cultural importance that the Fremen place on the conservation of water. The danger that Halleck responds to centres on the erosion of the discipline and social cohesion that had served as a touchstone for a cultural identity tied to the specificities of the environment, helping to bind the Fremen into a tight-knit global society.

The transformation of the environment and Fremen culture presented in *Children of Dune* is anticipated in *Dune Messiah*. Paul reflects on the changes that he initiates as the new Emperor:

> They hated him. He'd slain the past. And there were others, even those who'd fought for the sols to buy precious water, who hated him for changing the old ways. As the ecological pattern dictated by Muad'dib remade the planet's landscape, human resistance increased. Was it not presumptuous, he wondered, to think he could make over an entire planet – everything growing where and how he told it to grow? Even if he succeeded, what of the universe waiting out there? Did it fear similar treatment? (Herbert, 1971, 34)

The 'shortening of the way,' the rapid change to the environment and the social conditions that are constrained by the landscape, is the source of a profound deterritorialisation that is evident to Paul even as he makes these changes. Awareness of the social tensions that he precipitates leads him to abdicate as *Dune Messiah* concludes and retreat into the desert to adopt the role of prophet in *Children of Dune*. In this capacity he stands as a pastoral counterforce opposing the changes that he had himself initiated by establishing Arrakeen as the centre of a new civilisation. The heroic individualism and Machiavellian political self-interest that dominated *Dune* are here profoundly critiqued for their hubris and their extension outward to the universe. Paul notes that Arrakis itself 'fought him, resisted, slipped away from his commands,' illustrating how the nature of the planet resists the imposition of an individual's landscaping vision (Herbert, 1971, 56). The *Dune* trilogy presents this hubris as inescapable, eventually leading to Leto II's transcendent synthesis with the sandworms and complete control over the destiny of the Fremen and the planet. The *Dune* trilogy borrows from ecology and the pastoral to sketch a movement from terraforming seen as a positive physical and socio-cultural transformative force to terraforming as a problematic symbol of ecopolitical dictatorship and deterritorialisation.

Robert Heinlein's *The Moon Is a Harsh Mistress*

In Heinlein's *The Moon Is a Harsh Mistress,* Luna (the Moon) has been colonised by convicts deported from Earth generations ago, who labour to satisfy the increasing demand for resources on their home planet. This economic arrangement is overseen by the Authority, which represents Earth's interests on Luna and maintains a flow of cheap grain to the imperial centre. The narrative depicts a conspiracy by a group of 'rational anarchists' to unite the people of Luna, overthrow the Authority and declare independence from Earth. This text extends the political enquiry of *Farmer in the Sky* by examining how libertarianism offers an alternative to the imperial and fascist politics seen in the dystopias of the 1950s terraforming boom. The revolutionary Professor Bernardo de la Paz exhorts Earth's politicians to '[s]end us your poor, your dispossessed, send them by thousands and hundreds of thousands; we'll teach them swift, efficient Lunar methods of tunnel farming and ship you back unbelievable tonnage. Gentlemen, Luna is one enormous fallow farm, four thousand million hectares, waiting to be plowed!' (Heinlein, 2001, 240–41). The political context, with its reference to the American War of Independence and the colonisation of Van Dieman's island, invokes ideals of self-sufficiency and ecologically informed economics. The poor and dispossessed are relied on to abandon their lives on Earth to become colonial farmers; poised upon this injunction is an ambivalence between utopian hope for a better life on Luna and the potential exploitation of the workforce.

Geoengineering themes are explored in the context of an overpopulated and politically strained Earth trapped in a negative feedback cycle. The increasing demand for resources on Earth informs the Authority's use of criminals as slave labour on Luna, establishing a terragouging model of terraforming as the primary human relationship to Earth and the Moon's environment. At the narrative's conclusion, the assumption of a terragouging model for approaching other planets is not significantly challenged. As Luna settles into independence, Mannie considers the terraforming of the asteroids in terms that draw on space as a field of continued conquest, adventure and excitement (Heinlein, 2001, 303). The struggle for personal fulfilment outweighs the interest that development of Lunarian society holds for the text, which eventually turns outward towards space rather than inward to explore equitable social relationships on both Earth and Luna.

The Moon Is a Harsh Mistress bears many similarities to the consensus futures of the 1950s terraforming narratives. In order to combat Earth's exploitation of the Lunarian society for the production of cheap resources, de la Paz argues for self-sufficiency as a step towards a capitalist-libertarian ethic of the free market:

> Every load you ship to Terra condemns your grandchildren to slow death. The miracle of photosynthesis, the plant-and-animal cycle, is a closed cycle. You have opened it – and your lifeblood runs downhill to Terra. You don't need higher prices, one cannot eat money! What you need, what we all need, is an end to this loss. Embargo, utter and absolute. Luna must be self-sufficient! (Heinlein, 2001, 17)

De la Paz reprises Miller's symbol in 'Crucifixus Etiam' of the wine-soaked Martian sand, an emblem for the lives of the labourers that have been sapped by an alien environment. This image trades on the connection between ecological energy flows and blood, describing a feedback loop between the colonists and the land. De la Paz's call for a free market is grounded in sensitivity to Luna's limits: the scarcity of nutrients and water for cultivation makes its ecology exceedingly fragile. When this scarcity is coupled with the Malthusian population explosion that leaves many countries on Earth (India is the main example) unable to provide living space and food for their populace, the urgency of establishing a free market is intensified. Mannie's friend and co-conspirator, the AI known as Mike, uses a series of statistical scenarios to project a decline on Luna in seven years. De la Paz announces to the Lunar Authority on Earth that '[d]iscussions of how to augment our shipments must be based on the facts of nature, not on the false assumption that we are slaves, bound by a work quota we never made' (Heinlein, 2001, 83–84). The urgency of this scenario of ecological decline leads to the necessity of establishing economic, and hence political, self-determination on Luna.

This call for a scientific outlook that can appreciate and inform decisions based on the facts of energy economics is rooted in de la Paz's libertarian political philosophy, 'rational anarchism.' The political ends that discourses of nature are deployed to justify, Earth's instrumental relationship to Luna, are contested by a philosophy that privileges a certain value of rationalism. Jason Bourget has argued that Heinlein's reformulation of nineteenth-century libertarian political philosophy is undercut by a biological determinism that privileges masculinity, thus preventing the realisation of a libertarian utopia in the text (Bourget, 2008, 10–11). This is coupled to an insistence that politics suitable to the level of individual interaction can be scaled up to best equip a global society for survival. Rational anarchism locates social responsibility with a specific individual:

> A rational anarchist believes that concepts such as 'state' and 'society' and 'government' have no existence save as physically exemplified in the acts of self-responsible individuals. He believes that it is impossible to shift blame, share blame, distribute blame … as blame, guilt, responsibility are matters taking place inside human beings singly and nowhere else. But being rational, he knows that not all individuals hold his evaluations, so he tries to live perfectly in an imperfect world … aware that his effort will be less than perfect yet undismayed by self-knowledge of self-failure. (Heinlein, 2001, 51)

By coupling this political philosophy to notions of biological determinism, Bourget argues, 'Heinlein's populist revolution is rapidly transformed into an elitist dictatorship, dominated by a few charismatic men who assume complete political and economic control as soon as the Lunar Authority no longer exists' (Bourget, 2008, 18). De la Paz is one such charismatic figure, whom 'Heinlein unwittingly transforms […] from a representative of libertarian thought into a tyrant with a fit belief in the importance of his own masculine individuality' (19). The masculinist fascism that de la Paz represents recapitulates the Machiavellianism of Paul and Leto II in the *Dune* series, thus sustaining a tradition of instrumental relationships towards planetary environments and their inhabitants. In *The Moon Is a Harsh Mistress*, the deployment of grain shipments and rocks as projectiles to threaten Earth is one manifestation of this individualism and instrumentalism.

Rational anarchism represents a backlash against the idea of global governmental systems or, in Anderson's case, an interplanetary UN. It rejects bureaucratic systems and denies a global sense of identity rooted in responsibility to the group. However, as Bourget has shown, this individualism is restricted to those who see it as their duty to ensure their position as leaders directing the course of history. *The Moon Is a Harsh Mistress* closes spaces for a libertarianism that allows multiple voices to interact, which would open a dialogue about the meaning and function of a plural community: Mannie discovers that Mike has rigged the results of the election for the Lunarian council, giving Mannie a seat for which he had not been democratically elected. De la Paz calls this anarchism 'freedom,' a notion that Le Guin revises as a freedom to do anything, not a freedom from anything (Le Guin, 1999, 108). Heinlein's treatment of the terraformed Moon as an alternative, anarchist political system clashes with the implications of his biological determinism, but it does represent a departure from earlier ideas of global or interplanetary governments, and it responds to the same countercultural opposition that influenced the New Wave of sf.

Ursula K. Le Guin's *The Dispossessed*

Le Guin's *The Dispossessed* represents an alternative to the masculine politics of Herbert and Heinlein. Published almost a decade after *Dune* and *The Moon Is a Harsh Mistress, The Dispossessed* is informed by the successes and failures of the 1960s counterculture. Le Guin's interest in ecology, feminism and colonialism marks her as an early ecofeminist, while her work greatly influenced sf discourse and provided another voice that later writers would inherit. Fredric Jameson sees 'world reduction' as an essential strategy of Le Guin's work, which he explains is 'based on a principle of systematic exclusion, a kind of surgical excision of empirical reality, something like a process of ontological attenuation in which the sheer teeming multiplicity of what exists, of what we call reality, is deliberately thinned and weeded out through an operation of radical abstraction and simplification' (2005, 271). Darko Suvin summarises Jameson's reading of scarcity in *The Dispossessed* as 'a reaction to the polluted American abundance and a realistic diagnosis of a better model of life,' and extends this claim to argue that world reduction is 'a retrenchment from the "living flesh" of a natural community, a harsh but clean acceptance of asceticism' (1975, 273). Ursula K. Heise, however, argues that Jameson reduces 'Le Guin's carefully conceived planetary ecosystems [to] a purely metaphorical function,' thus overlooking the way the ecological constraints of Anarres '[shape] social and political horizons in particular – both positive and negative – ways' (2012, 101, 103). Gib Prettyman likewise observes that Jameson and Suvin ignore Le Guin's use of ecology and Daoism, 'both frameworks [which] emphasize systemic processes and aim to critique egoistic illusions' (2014, 58).

The Dispossessed makes much of the implications of its subtitle and status as an 'ambiguous' or 'critical utopia,' in Tom Moylan's theorisation (1986). The terraforming tradition taken as a whole operates analogously to the critical utopia through techniques such as the dialogising relationship established between *The Moon Is a Harsh Mistress* and *The Dispossessed*. Alternating chapters set primarily on the planet Urras and Anarres allow readers to compare their political systems as seen from the protagonist's perspective. While Shevek, the protagonist of *The Dispossessed*, contrasts Urras against the touchstone of his home planet, Anarres, the narrative's structure dialogises this with scenes from Shevek's past, thus disclosing flaws in the Odonian political system. This narrative strategy contributes to the ambiguity of the text's utopianism. Pastoral themes differ from world reduction as, rather than excising elements of empirical reality, the dialogism and compression of meaning involved in the pastoral emblem encode a wider range of experience and generate an allusive network that opens the text to complexity. Pastoral emblems, while a simplification of this complexity, operate by compression rather than exclusion. Working in concert with these textual strategies is the mobilisation of ecological themes that limit the development of Anarresti society and which, as Heise argues, '[hold] up environmentalist discourses of scarcity as they had emerged in the 1960s and early 1970s to the reader's scrutiny as conceptual tools whose usefulness in shaping contemporary societies' relation to their environments needs to be constantly re-examined' (2012, 103).

Anarres is an anarchist commune, a social experiment in anarchist political philosophy. Odonian government is designed to escape the shortcomings that have resulted in an uneven set of social relationships on Urras. In contrast, the environmental concerns of A-Io are genuine but socially stratified, primarily working to maintain the economically privileged in a state of luxury. Decentralisation is combined with the Odonians' repudiation of tribalism in the name of a broader conception of community. Tribalism, however, is extended to the Odonians as a whole: many reject Shevek's campaign for contact with Urras, preferring instead to operate as a closed, self-sustaining commune. Bramwell argues that 'religious minority groups, such as the Amish and the Doukhoubours [*sic*], survived because they were transplanted as homogenous groups from their country of origin, and owed their survival to tribal as well as religious bonds, rather than novel experiments' (1990, 93). Such homogenous transplanting is also characteristic of the foundation of Odonian society, yet separation is already compromised by the fact that Anarres is an Urrasti mining colony (Le Guin, 2000, 79). While Anarresti tribalism draws identity and collective strength from a political text written by the spiritual founder Odo, Shevek and his supporters see this collectivism as having rigidified into a restraint.

One crucial ecopolitical element is the pastoral inversion between Urras and Anarres. The pastoral world and its pairing with capitalist economics and a network of international relations on Urras abstract contemporaneous real-world perspectives on global politics. This is complicated when Shevek discovers that he has been insulated from the oppression of the urban experience of the non-elites in A-Io. Recalling Ernest J. Yanarella's discussion of the Garden of the Chattel form of pastoral, this concealment parallels the pastoral ideal's use as a rhetorical strategy that cloaks or elides the necessities of everyday life and the dynamics of oppression (2001, 81–82). Shevek's experience of the landscape and animals of Urras broadens this pastoral theme by including vectors for the recognition of nature's otherness. Shevek's first sight of a horse and a flock of birds are haunting images; his experience with animals in A-Io enacts a confrontation with the alien but establishes this event as a reconnection. These visions delineate a perspective that allows Shevek to reassess his view of Urras and entertain the possibility of a human community spanning the two societies.

After Shevek accepts a posting where he works to plant vegetation engineered for environments on Anarres, he becomes acquainted with a pastoral song sung by the labourers. His reaction to some of its oblique references underscores the otherness of the pastoral landscape to the Odonians. Nevertheless, for many of the labourers the song retains an inspirational element that underwrites their efforts to terraform their moon. The name of the first site on Anarres, Ans Hos, can be translated from the Odonian language Pravic into the Ioti of A-Io as 'Garden of Mind', and the site is known as the Eden of Anarres. The chronotopicity of this space, as a colonial settlement set in an unspecified future, is counterpoised with the pastoral nostalgia that it evokes. The physical constraints that Anarres poses divert attention away from the instantiation of the pastoral landscape and towards the psychical processes of landscaping and imagining the future.

Odonian society avoids the use of unnecessary technology but is not primitivist: they actively reject this pastoral orientation towards an idealised past in favour of a 'complex organicism' (Le Guin, 2000, 81). The combination of decentralisation and a low-tech infrastructure, the latter of which recalls the feudal setting in *Dune*, reflects a distrust of large, bureaucratic governments where decision making devolves to an imperial centre. Nevertheless, there are centres on Anarres: the centralised computer system Divlab manages the planet's work postings, while the capital Abbenay is a place where power inheres. This ironically undercuts Shevek's first impression that, in Abbenay, nothing is hidden. Anarres and Urras are unveiled during the course of the narrative while the Odonians' anarchism is critically examined, its flaws identified, and a restoration of its basic principles attempted.

Such contradictions exemplify how *The Dispossessed* works as an ambiguous utopia that undermines the ideal. Nevertheless, the differences between the two governmental and philosophical systems are real. Odonian social philosophy, based on Kropotkin's concept of 'mutual aid' and a distrust of rigid social practice, is a significant indicator of a shift in political philosophy in sf that influenced the ecopolitical dimension of terraforming narratives such as Kim Stanley Robinson's *Mars* trilogy. Anarchism, as Shevek argues, is part of their 'nature' as Odonians. In contrast, Chifoilisk, a scholar on Urras, offers another assessment of human nature that is clearly representative of those of his society. In one episode set in Shevek's youth, he and his friends look up at the distant

Urras and see a beautiful cosmic body. Tirin suggests that it might be desirable to see what is happening on Urras, arguing that at the very least it would be worth going to see a horse. Despite Shevek's scepticism at what he believes to be a childish notion, it speaks of a desire to reach beyond an immediate community to expand ideas of a lived sense of place from global to interplanetary scales. Tirin expresses a desire to reconnect with the other so as to re-evaluate the immediate community from an external perspective and to open spaces for overtures towards a positive recreation of Odonian society.

This re-evaluation of anarchism turns on what Shevek calls the 'nature of existence' and on notions of brotherhood (Le Guin, 2000, 52). Against the popular Odonian belief that brotherhood begins with love, Shevek contests the idea that suffering is simply a social disease, arguing that brotherhood begins in shared suffering – the very thing another Odonian argues that mutual aid is designed to prevent. While society can work to eliminate unnecessary suffering, Shevek argues that it is part of the nature of existence. Although his argument is questionable on the ground that there are other ineliminable traits that remain part of the nature of existence (one might argue that less commendable human qualities could also have equal claim to centrality, as does Chifoilisk), the connection between suffering and brotherhood makes this trait significant for its social and ecopolitical implications.

In light of this exchange, Shevek's call of brotherhood to a pet otter is a call for solidarity that draws connections between ideas of shared suffering between distinct species. The common factor linking their experience is not their embodiment, but the suffering caused by shared environmental conditions. Their historical relationship as creatures that had formerly occupied the same world as the Odonians gestures towards the possibility of re-territorialisation through an animal other. Shevek's experience with animals is combined with a thread of moral extensionism that balances his sense of nature's otherness with a solidarity based on recognising his affinity with other animals. This element of moral extensionism soon overrides nature's otherness and is linked to the anarchist ideals explored in human political contexts. Shevek grounds human relationships to other animals and their environments in the same political philosophy as his view of intra-human relationships.

Shevek shames one scholar, Pae, by showing contempt for the comparison he draws between the laws of physics and the uneven political arrangements on Urras. Mathematics undergirds the laws of physics, while thematic similarities between ecology and physics highlight the coherence between their emphasis on ideas of unification and connection in their account of different aspects of nature. Mathematics, or 'number,' is a bridge between psyche and matter. Number is the basis of other modes, and Ainsetain (Shevek believes) thought of physics and mathematics as an accurate description of reality. If Shevek's understanding of number is correct, this basis in reality is the ground where all political and ethical systems are rooted. Shevek's mission to discover a unified field theory grows from the same root as his desire to establish a connection between Anarres and Urras. Connections underlie Shevek's continued interest in social reform as one example of his attempt to unify separate domains. He makes this clear when claiming that reconciling Simultaneity and Sequency would amount to a complexity that embraced both geometry and ethics.

Reconciling Simultaneity and Sequency recapitulates the Daoist framework that is central to Le Guin's oeuvre, apparent in the recurrent theme of balance between opposites. Prettyman notes that while world reduction in *The Dispossessed* does attempt, as Jameson suggests, to imagine capitalism away, 'it is important to notice that [Le Guin] saw her yin utopianism as a strategic counterweight rather than a mystical celebration of inevitable balance' (2014, 63). In this light, the difference in emphasis on love and violence in 'Vaster Than Empires and More Slow' and *The Word for World Is Forest* can be read as an attempt to use the Hainish universe to explore multiple aspects of balance in society. Instead of 'prevailing egoistic orientation[s] toward endless growth,' Le Guin 'wants to emphasize a radical knowledge of place, of here and now' (Prettyman, 2014, 65). Shevek's dynamic interrelation of Simultaneity and Sequency gives prominence to the present and its connection to the past and future, and is complemented by the ecological themes that foreground issues of species flourishing and space. Shevek's partner Takver is a biologist who specialises in fish genetics. Shevek sees Takver's 'concern with landscapes and living creatures' as something 'much broader than love,' and speaks of her as being unweaned from the universe (Le Guin, 2000, 154). As a counterbalance to an egoism that privileges individuality, Takver's relationship to the timespace of the now involves a reciprocal expansion of identity that retains distinctions while recognising continuity: '[i]t was strange to see Takver take a leaf into her hand, or even a rock. She became an extension of it: it of her' (154).

Shevek turns to the Terrans and Hainish for asylum after the brutal suppression of the Ioti rebellion. The Terran ambassador Keng tells Shevek of her homeworld (Earth), which was devastated by the inhabitants' inability to adapt to the environmental conditions created by their multiplying population and practices. To Keng, Urras, despite its faults, is a world tremendously alive. This contrast appears to mitigate, in Keng's eyes at least, the suppression of the rebellion. Shevek, on the other hand, connects this viewpoint to his own understanding of time and to his work as a physicist. Speaking from a position informed by the insights of Simultaneity, Shevek argues that you cannot have the present unless you accept both the past and the existence of the future. This dynamic is, as Moylan argues, typical of *The Dispossessed*, which he argues expresses an attitude of *détente*, 'the cooperation of previously contending forces to transcend hostility, suffering, and injustice and work jointly toward a better world for all' (1986, 93). From the perspective of Sequency, Shevek assesses Keng's nihilism as an inability to accept change, to accept that evolution does influence both human nature and culture. This extension of physics into the domain of evolution and ethics points towards a utopian future that is ambiguous because it cannot exist in static form. Instead, much like Shevek's view of Odonian society as rebellious, the process of evolution is paralleled with that of revolution, both of which are ongoing and permanent.

This process of revolution is first and foremost an individual act, a condition that is reflected by the narrative's focalisation through Shevek, one dimension of anarchism that sits uncomfortably with ideas of community and survival that terraforming traditionally emphasises. As Shevek argues – faintly echoing Poul Anderson's anti-collectivism – 'any rule is tyranny,' including the rule of the group, and if an individual cannot work in solidarity with their community then their duty is to work alone (Le Guin, 2000, 295). The problem of the individual and the group rises to the fore as the main conflict within Anarresti society and remains troublesome for conceptions of anarchism. The theme of heroic individualism, often associated with imperial and patriarchal conceptions of right government, is refigured in less heroic terms and presented as a necessary step towards social change and responsibility. Shevek's brand of individualism differs from conceptions of heroic individualism in that it is balanced by a strong conception of brotherhood as a binding social force.

Shevek reflects on fulfilment as a function of time and uses two conceptual chronotopes, the motif of the locked room and the landscape of time outside this metaphorical locked room. The room is a symbol for individual stasis and intellectual imprisonment and appears throughout the text in various forms, most notably as the motif of the wall that opens the narrative. The idea that an act becomes human only when it occurs in the landscape of time, in both the past and the future, embeds the individual into their immediate environment and emphasises the dynamism that undergirds physics and ecology as the basis for ethics and culture. Loyalty ties the past and the future together, establishing continuity between temporal landscapes – between interpretations of the past and visions of the future. Terraforming is both physical and social, involving a superimposition of physical and intellectual landscapes. Conceptions of time are intellectual landscapes, while loyalty is a political relationship that fuses the physical aspects of terraforming to a politics oriented to the future. Reflection on the landscape of the past and an acceptance of change, both physical (the terraformation of planets) and social (the reformation of socio-political landscapes), are brought together in the terraforming text. Such narratives provide spaces where the physical and political are drawn together; the terraforming motif operates as an emblem for the inter-animation of these two domains.

Ernest Callenbach's *Ecotopia*

Ernest Callenbach's *Ecotopia* (1975) is a classic of ecological literature. Andrew G. Kirk writes that '*Ecotopia* melded the counter-culture lifestyle and social values with a strange brew of libertarian thinking, collectivism, states rights, and technologically enthusiastic environmentalism in the same counterculture sci-fi tradition as Robert Heinlein's *The Moon Is a Harsh Mistress*' (2007, 157). *Ecotopia* operates more like a catalogue of already existing technologies and an exploration of how they could be utilised within a particular social framework, and in this sense it resembles Stewart Brand's *Whole Earth Catalog*. Indeed, in 1972 Callenbach presented the ideas that would feed into *Ecotopia* to the Point Foundation committee, which was founded by Brand and funded by proceeds from the *Whole Earth Catalog* and the *CoEvolution Quarterly*. This presentation was itself a work of performance sf, as Callenbach assumed the role of an 'Ecotopian Militant' from 1995 to explore ideas of a decentralised economy (136). *Ecotopia* links the countercultural ecologism of the American West Coast, with its exploration of science and technology as a way to synthesise nature and culture in terms recalling the motif of the machine in the garden, to the contributions sf makes in exploring what such a synthesis might look like.

Callenbach locates Ecotopia at the American West Coast, thus helping to establish an ecological and egalitarian tradition of green utopias to which Le Guin and Robinson contribute. *Ecotopia* is part of a discourse of ecologism that connects utopianism with eco-libertarianism and westward expansion: in Robinson's *The Gold Coast*, one character argues that 'Orange County is the end of history, its purest product. Civilization kept moving west for thousands of years, in a sunset tropism, until they came to the edge here on the Pacific and they couldn't go any farther. And so they stopped here and *did it*': they attempted to create a utopia (1989, 3). Californian ecological sf is concerned with connecting history to the future, opening the ecological utopian symbol of Orange County to ideas that coalesced around the *Whole Earth Catalog* and its related publications: ideas of ecological design, appropriate technologies, a libertarian distrust of centralised bureaucracies and a desire for social freedom and free markets.

The influence of the term 'Ecotopia' on sf is evidenced by Robinson's publication of an anthology of short stories titled *Future Primitive: The New Ecotopias* (1997). *Ecotopia* begins with an epigraph that translates the etymology of the term 'ecotopia': the root 'Eco-,' 'from the Greek *oikos* (household or home),' highlights the etymological connection between utopia and ecologism (Callenbach, 1978, n.). This signification is central to terraforming, given its concern with constructing new homes on other planets. Le Guin, too, draws on the relationship between ecologism, utopia and home, most notably in her experimental critical ecotopia *Always Coming Home* (1985). Werner Christie Mathisen sees the connotations of home as implying 'an exemption from difficult choices, and thereby a reduction in individual and social autonomy,' although Le Guin and Robinson have consistently challenged such abdication of personal choice in their narratives (2001, 69). Terraforming, as a literature of landscaping and habitation, makes ecology central to the realisation of a critical utopian space in the late 1970s, thus factoring the more-than-human world into its exploration of social reconstruction. These narratives can thus be considered works that express an eco-cosmopolitan vision.

Ecotopia deals with geoengineering themes that are linked to the governmental and economic restructuring of society. Its secessionist politics represents a formal link to the relationship between Earth and other terraformed planets. Separation from America allows the Ecotopian society to develop according to its own socio-political trajectory, a dynamic that is magnified in the spatial separation between Earth and terraformed worlds. Some of the main elements of the Ecotopian political vision include an emphasis on decentralisation and stable state economics, which impact upon multiple levels of Ecotopian life, including production, education, local and regional politics, healthcare and town planning. A battery of lifestyle changes are presented, all of which index a shift of values and conceptions of identity. Ecotopian lifestyle resembles the hippie movement of the 1960s, while the neo-pagan practice of worshipping trees, which draws from the influence of Lovelock's Gaia hypothesis (Kirk, 2007, 169), illustrates the consolidation of a series of 'green' philosophies into an identifiable discourse.

The uneasy balance between a distrust of technology and its use, while grounded in fears of mechanisation and depersonalisation, counters the notion that the Ecotopians are atechnological primitivists. They tend to view technology as a support structure to facilitate a return to the wilderness. The narrator, Weston, notes that the Ecotopians' 'technological austerity' is belied by their extensive use of telecommunications, trains and sophisticated and innovative methods of generating power (Callenbach, 1978, 38, 68). This use of technology makes possible the arrangement of a collective life in a decentralised government system. The notion of stable state economics, which underlies theories of the sustainability movement, is taken as a blueprint that structures policy making at every level. Reaching the goal of a stable state involves decentralisation processes that, it is believed, will put humanity in a less destructive and exploitative relationship to nature by assisting in maintaining nature's ecological integrity. It is a homeostatic principle that has a corresponding feedback effect on social worlds. Stable states are also central to the notion of terraforming, the ideal of which is to establish a contained self-sustaining system. While Mathisen argues that *Ecotopia* and *The Dispossessed* consistently underestimate the role of politics in green utopias (Mathisen, 2001), they represent an important development of the terraforming narrative where critical reflection and ambiguity – as exemplified by *The Dispossessed* and the confluence of egalitarian politics and ecology – are instantiated.

The narratives explored in this chapter track a fundamental transformation of the way relationships to the environment and politics are envisioned in sf. Two related types of narrative co-exist and mutually support each other: proto-Gaian stories and terraforming narratives that develop their enquiry around principles influenced by energy economics. Both types of narrative challenge the assumptions of the works of the 1950s terraforming boom and extend the character of their ecopolitical enquiry. Oppositional politics rises to the fore as the terraforming motif is used to experiment with alternative lifestyles and forms of government. Such engagement is both abstract, as in McKenna's and Le Guin's proto-Gaian stories, and concrete, as in Heinlein's and Le Guin's consideration of governmental systems and lifestyles. It is during this period that proto-Gaian and terraforming narratives begin to converge. Sf begins to develop a distinctly 'green' discourse, a greening of the terraforming narrative that is exemplified by the Californian ecologism represented by Le Guin and Callenbach, and which would be extended by other writers throughout the 1980s–1990s. The next chapter turns to this period to examine the social and political development of the terraforming narrative during a time in which Lovelock's Gaia hypothesis becomes incorporated into sf discourse to inform the imagination of planetary adaptation.

4: Edging Towards an Eco-cosmopolitan Vision

Ursula K. Heise argues for 'the urgency of developing an ideal of "eco-cosmopolitanism," or environmental world citizenship,' claiming that it is 'imperative to reorient current U.S. environmentalist discourse, ecocriticism included, toward a more nuanced understanding of how both local cultural and ecological systems are imbricated in global ones' (2008, 10, 59). She considers local, national and global forms of identity manifest in the environmental movement since the 1960s and in ecocriticism in the 1990s, tracing the shifting scholarly debate regarding conceptions of the local and global. Heise begins with the concept of globalisation, which rose to prominence in the late 1990s as 'the central term around which theories of current politics, society, and culture in the humanities and social sciences are organized' (4). Although the parameters of these debates and the poststructuralist critique of essentialist nation-based identities that arose began in the early 1980s to mid-1990s, Heise claims that globalisation is beginning to supersede others in theories of postmodernism and postcolonialism. Theories of hybridity, creolisation, *mestizaje*, migration, boderlands, diaspora, nomadism, exile and deterritorialisation provided countermodels to essentialist, nation-based concepts of identity, while in the later 1990s concepts such as 'transnationalism' and 'critical internationalism' and the resurgent concept of 'cosmopolitanism' in discussions of globalisation began to take priority. She observes that counter-critiques to globalisation have emphasised 'the value of local and national identities as forms of resistance to some dimensions of globalization,' resulting in a theoretical impasse (5–6). Against this scholarly background, Heise argues that modern environmentalism has been concerned with issues of the local and global since the movements of the 1960s and 1970s.

Terraforming stories have explored concerns that Heise argues are central to an eco-cosmopolitan awareness, having considered the politics of globalisation since Wells's *The Shape of Things to Come*. Heise's discussion of deterritorialisation offers a useful theoretical concept for organising the themes explored in preceding chapters of this book. As terraforming is a process of adaptation and habitation, it too can lead to processes of deterritorialisation. Heise explains that

> The increasing connectedness of societies around the globe entails the emergence of new forms of culture that are no longer anchored in place, in a process that many theorists have referred to as 'deterritorialization.' Undoubtedly, deterritorialization, especially when it is imposed from outside, is sometimes accompanied by experiences of loss, deprivation, or disenfranchisement [...] yet deterritorialization also implies possibilities for new cultural encounters and a broadening of horizons that environmentalists as well as other politically progressive movements have welcomed, sometimes without fully acknowledging the entanglements of such cultural unfolding with globalization processes that they otherwise reject. (Heise, 2008, 10)

Sf has portrayed responses to the estranging dynamic underlying deterritorialisation in terms of a reactionary refusal of the radically other or a celebration of the alien that leads to a re-valuation of the social, political, cultural or cognitive landscape. Encounters with cosmological nature's otherness, with alien ecologies, cultures and alternative socio-political arrangements on planets undergoing terraforming, explore processes of deterritorialisation and re-territorialisation in the colonists' experiments with the possibilities for developing new attachments to alien landscapes.

Terraforming foregrounds a range of heterogeneous spaces, connecting them in ways that parallel the dynamic Heise identifies as the use of 'collage' or 'montage' techniques in literature (2008). Heise's analysis focuses on several works of environmental sf, including Gaian stories such as Ursula K. Le Guin's 'Vaster Than Empires and More Slow' (1982 [1971]) and David Brin's *Earth* (1990). She argues that these works 'attempt to develop aesthetic forms that do justice both to the sense that places are inexorably connected to the planet as a whole and to the perception that this wholeness encompasses vast heterogeneities by imagining the global environment as a kind of collage in which all the parts are connected but also lead lives of their own' (Heise, 2008, 64). The portrayal of spaces associated with the colonising and terraforming project, the pastoral chronotopes of the farm, garden and wilderness, along with spaces such as forests and seas, develops a montage effect akin to this collage technique. Individual chronotopes are often overlaid with multiple landscapes that correspond to alternative orientations to nature. Narratives of terraforming build worlds containing various global and local chronotopes, contrasting and connecting them in ways that construct new relationships for negotiating ecopolitical and ecophilosophical issues.

The publication of James Lovelock and Michael Allaby's *The Greening of Mars* in 1984 signals the explicit convergence of terraforming and Gaian themes in sf. *The Greening of Mars* shares many stylistic features with Ernest Callenbach's *Ecotopia* (1978 [1975]) but focuses on the scientific principles substrating terraforming and on propagandising for interplanetary colonisation. Like Callenbach, both Lovelock and Lynn Margulis wrote in the 1970s for the *CoEvolution Quarterly,* a publication launched from the proceeds of the *Whole Earth Catalog.* It was through publications such as this and Lovelock's popularising scientific works, beginning with *Gaia* in 1979, that the Gaia hypothesis was first introduced to the public. Combining ecotopian speculation with a tradition of hard sf, Allaby and Lovelock use the terraforming narrative as a vehicle for popularising theories about Gaia (1984).

The Greening of Mars argues that terraforming is not only politically desirable but feasible with the technology of the 1980s, albeit only as a private undertaking. Combining technological methods for terraforming Mars with a model derived from the assumptions of the Gaia hypothesis, '[t]he idea of "greening" Mars', says the narrator, 'rather than "industrializing" it was inherently attractive. It seemed gentler, more "natural," and it was more natural too, in that the transformation was to be achieved by the activities of living organisms, left to their own devices' (Allaby and Lovelock, 1984, 112–13). Developing the links between terraforming and Gaia offers romantic subject matter for the exposition of the relationship between science and geopolitics. As the narrator explains, this requires that he 'discuss the nature of life itself, and the ways in which it manifests itself' (49).

It was at a scientific meeting inspired by *The Greening of Mars* that Robert H. Haynes coined 'ecopoiesis' – from the Greek root *oikos* and *poiesis,* meaning creation, production and fabrication, literally 'the making of a home' – for this process of planetary adaptation (1990, 161–83). This term has circulated widely in terraforming stories since the 1990s, appearing in Frederick Turner's epic poem *Genesis* (1988), Frederik Pohl's *Mining the Oort* (1992), NASA scientist Geoffrey A. Landis's short story 'Ecopoesis' (2001 [1994]) and Kim Stanley Robinson's *Mars* trilogy (1996c [1992]; 1996b [1993]; 1996a). Ecopoiesis is an approach to landscaping that connects terraforming as fabrication to a biotic process aligned with craft-work – the shaping of ecosystems – that works to limit both the deterritorialisation involved in living on another planet and the alienation from nature that is associated with industrial methods for adapting the environment.

The Greening of Mars purports to be an account of the history of Mars colonisation, written during the narrator's return to the planet. It is interspersed with lengthy historical and scientific discussion of a range of subjects related to colonising Mars and connects political and scientific ideas about migration. The decision to proceed with an ecopoietic model of planetary adaptation is political, because it would allow the colonists to develop economic independence from Earth. As the narrator explains, '[t]here might be industrialists, but they would be martian [*sic*] industrialists – the colonists themselves' (Allaby and Lovelock, 1984, 114). This emphasis on economic and political autonomy is linked to evolution and species survival: the narrator describes Martian independence as an answer to the cultural ossification caused by private companies and colonial governments. Independence promotes flexibility towards changing conditions, a trait essential to the physical evolution of species in periods of environmental change.

The narrator points to an analogy between emigration from Europe and interplanetary migration, noting that in both cases colonists 'cease to be part of the continuing history of their original country or planet' (Allaby and Lovelock, 1984, 114). Nevertheless, the narrator is careful to assert that such analogies fail to adequately explain Mars's history and are incompatible with the realities of colonising the planet. At the climax of the narrative, the narrator reflects on the diverging histories of humanity on Earth and Mars. The purpose of the narrator's visit to Earth is to deliver a scientific report for verification by Terran scientists. This report describes the speciation of the Martian colonists and asserts that 'we Martians now comprise a distinct species within the genus *Homo*' (154). Divergence from the Terran genetic code makes literal the divergence of human history that Mars colonisation entails. Physical and cultural dissociation from Earth's history is part of a process of re-territorialisation where the colonists, under the impact of the 'radically different' Martian environment, labour to make Mars a home (160). Taking evolution as the determining factor for human physical and cultural development, which are treated as 'biology, nothing more,' this divergence from Earth's history is both a disturbing and a challenging phenomenon that carries with it the potential for 'real improvements' that should be welcomed (163).

Divergence opens up the threat of conflict between human groups. This is complicated by the differences between the elite Martian colonists – those who have inhabited Mars for generations – and new arrivals who have not yet adapted to the environment. The narrator raises fears that a form of racialism may emerge, one that would magnify the dynamic of conflict between cultural and ethnic groups on Earth. Mars's elite 'developed its own customs and, more important, its own aesthetic concepts. Its members felt they had, to a large extent, "built" Mars' (154). The development of new aesthetic categories that have repercussions for Martian lifestyles is a sign that the colonists are inventing new relationships with the landscape. The colonists' decision to opt for an ecopoietic model of planetary adaptation in contrast to an industrial mode is reflected by the specific sense in which Mars is seen as having been 'built': 'Mars could have been rebuilt, physically and chemically, as a replica of Earth, rather than being encouraged to develop in its own ways' (109). The pastoralism of ecopoiesis ameliorates the anxiety associated with inventing novel cultures and lifestyles. They are themselves fabrications in the sense Tom Shippey alludes to in his contrast between pastoral and fabril literatures. *The Greening of Mars* attempts to demystify many of the romantic assumptions embedded in the notion of terraforming as replicating Earth, using representations of the physical adaptation of Mars as a way to tackle questions of migration and deterritorialisation.

Pamela Sargent's *Venus* trilogy (1989a [1986]; 1989b [1988]; 2001a) responds to the problem of creating a sense of place that would allow the colonists to develop new relations to each other and to their environments. Sargent's trilogy turns away from Venus and Earth by shifting focus from the development of suitable lived experiences within a dynamic and autonomous planetary environment. Sargent's *Venus* trilogy and Turner's epic poem *Genesis* represent two significantly different treatments of terraforming, despite sharing fundamental similarities that underlie attitudes towards planetary adaptation and their socio-political and enviro-ethical implications. This chapter considers how they have presented sophisticated engagements with ecopolitical issues to explore the possibility of developing an eco-cosmopolitan awareness. First, however, a discussion of Sargent's 'Dream of Venus' (2001b [2000]) illustrates the explicit focus on environmental philosophy and the ramified dialogism of the terraforming narratives of this period.

'Dream of Venus' was published after Kim Stanley Robinson's *Mars* trilogy, the last book of which appeared before the completion of Sargent's *Venus* trilogy. Although the critical and popular success of Robinson's work has overshadowed Sargent's, both engage in a sophisticated ecopolitical critique of society. 'Dream of Venus,' perhaps in response to Robinson's focus on environmental philosophy, but certainly in response to the burgeoning interest in this area since the 1990s, reflects explicitly on what was implicit in Sargent's earlier work: notions of intrinsic value, environmental aesthetics and respect for nature's otherness.

The geologist Hassan Petrovich Maksutov is charged with assisting Miriam Lucea-Noyes in creating a mind tour of Venus, a virtual reality entertainment which Donald M. Hassler suggests 'is an echo back to the pulp genre of SF itself' (1997, 155). Venus's administrator Pavel Gvishiani acknowledges that the mind tour is a masterpiece but, because of political restraints, requests that they censor it: Miriam refuses; Hassan agrees. Earth's government continues to influence the direction of the terraforming project, thus forcing Pavel to dismiss Miriam and order Hassan to edit the mind tour to remove its offending sections. Miriam and Hassan part and, years later, Pavel hints that he has preserved and distributed the mind tour amongst the linkers (the social elite), reminding Hassan of the dream he once shared with Miriam.

The device of the mind tour allows Sargent to consider questions of humankind's aesthetic response to nature: entitled *The Dream of Venus,* it depicts the natural history of Venus's landscape. Miriam's creative vision and her struggle with the administration's censorship suggest links between artistic creation, terraforming as the creation of new worlds and the instantiation of a civilisation's 'dreams.' It soon becomes apparent to Hassan that Miriam is a visionary artist whose creation offers an alternative to the orthodox future dreamed of by Earth's officials: '[s]omehow she had taken what could have been no more than an impressive visual panorama and had found the beauty in the strange, alien terrain of Venus as it might have been six hundred million years ago. It was as if she had fallen in love with that world, almost as if she regretted its loss' (Sargent, 2001b, 405). The sense of beauty that Hassan attributes to Miriam's depiction of Venus rests on the sense of its strangeness. Miriam's love for the planet and her efforts to communicate this feeling implicitly criticise the terraforming project, giving her artistic endeavour a political dimension that prompts the administration's suppression of the mind tour. This aesthetic response is first directed at a Venus displaced into the past, where, as some specialists argue, the planet's climate closely resembled Earth's. This oscillation between tendencies to familiarise the planet's alien otherness by appeal to proposed similarities with Earth and the recognition of the otherness of Venus gives *The Dream of Venus* its creative dynamic, allowing Miriam to shape Venus's history into 'a moving evocation of a planet's life, a depiction of a truly alien beauty' (Sargent, 2001b, 407).

This visionary love of otherness is politically destabilising for its implicit criticism of the Mukhtars' terraforming of Venus. Their initial desire for a propagandistic work to encourage faith in the project on both Earth and Venus admits a contesting voice that speaks on behalf of nature's autonomy. Shamed both by Pavel's distribution of the mind tour in opposition to the Council of Mukhtars and by his own betrayal of Miriam's love, Hassan experiences an epiphany regarding the fraught nature of political dialogue:

> the authentic dream, after all, was still alive. Dreams had clashed, he knew, and only one would prevail. But how would it win out? It would be the victory of one idea, as expressed in the final outcome of the Project, overlaid upon opposed realities that could not be wished away. To his surprise, these thoughts filled him with a calm, deep pleasure he had rarely felt in his life, and *The Dream of Venus* was alive again inside him for one brief moment of joy before he let it go. (Sargent, 2001b, 416)

Hassan's epiphany is Bakhtinian, and he sees his contribution to the Venus project as a small part of a still unresolved work, thus admitting the possibility that the dream he had foreclosed could still be realised. Latent voices underlie dominant ideologies, suggesting the possibility of their actualisation. Politics is seen as conflict, a clash between incompatible dreams that, because they are 'overlaid upon opposed realities,' are always defined by their opposition. This view is utopian, emphasising the unreality of the future and the work involved in its construction, which is first given shape by a vision of one future against others. Hassan, buoyed by joy and hope, turns away from what he feels is 'the authentic dream,' thus highlighting the difficulty of raising support for marginalised positions. The current political reality, however, is still in flux and unresolved, pointing to a future of hope that the *Venus* trilogy explores.

Building Critical Spaces: Pamela Sargent's *Venus* Trilogy

Sargent's *Venus* trilogy is set in Earth's far future, where the first planetary adaptations of Venus are located in a legendary past that acts as background for the portrayal of a vast engineering project conceived, for reasons of political control, to occupy the imaginative horizons of Earth's global society. Earth's Islamic government grew from a society that achieved political dominance after the Resource Wars devastated the planet. Its administration is divided into regional Nomarchies presided over by 'Mukhtars' – from Arabic *iktara*, meaning 'to select, choose' ('Mukhtar, n.,' 2015) – elites elected by a council that occupies itself with political intrigue and fears of revolt by a disaffected populace. The Mukhtars use the terraforming project as a way to deflect desire towards a grand social endeavour, 'a new dream, one that would inspire Earth's people' or, as is suggested, one that would defuse dissension that, unchecked, might threaten the Mukhtars' political dominance (Sargent, 2001a, x). Terraforming legitimises and maintains the Mukhtars' rule by providing society an outlet for directing their frustrations and energies, emotions rooted in the global society's rigid social stratification and collective desires.

Venus's extremely unforgiving physical characteristics present significant practical difficulties for the terraforming project, despite the existence of various technologies that allow the Nomarchies to manage Earth's climate. The terraforming project on Venus was instituted six centuries ago, shortly after the cessation of the Resource Wars by the legendary Mukhtar Karim al-Anwar, who by all accounts 'saw people who needed a new dream, a goal that might lift them to greater endeavors that would rival the accomplishments of the Associated Habitats and their people, who had abandoned Earth' (Sargent, 1989b, 11). Terraforming Venus is made possible only by the Mukhtars' reliance on the aid of the comparatively advanced 'Habbers,' humans who permanently reside in habitats in space. Reliance on a group that had historically rejected the Nomarchies and who subsequently function as the object of their propaganda compromises the technological supremacy that legitimises the Mukhtars' rule. Because the terraforming project is intended to operate as a monument to the Mukhtars' supremacy, the Habbers' aid necessitates (from the Mukhtars' point of view) a continued campaign of prejudice and disavowal. The Habbers, for their part, are the 'descendants of those Earthfolk who had abandoned a planet that they saw as a worn-out husk, who fled from the aftermath of the Resource Wars into space instead of staying to rebuild their damaged Earth' (Sargent, 2001a, x). Sargent thus calls these figures 'cyberutopians rather than cyberpunks' (Engel, 1991, 21). They reject terraforming and planetary habitation on moral grounds and because of a complex and more diffuse association between political and imaginative constraint and the bounded nature of planets, in contrast to the horizonless reaches and purity associated with the chronotopicity of space.

Robinson's *Mars* trilogy locates terraformation in the near future, in the politico-economic context of an Earth that closely resembles contemporary social relations. Mars's comparative hospitability makes it a more likely candidate than Venus for near-future terraforming. It is a chronotope that offers a significant narrative contrast to the complex of socio-political and economic factors constructed in Sargent's

Venus trilogy. Significantly, Sargent denies the narrative potential that terraforming Mars offers by having the Habbers claim it in the historical past of the *Venus* trilogy as a site of preservation against the human modification of its natural planetary environment. This general outline serves to delineate a range of themes that are central to these extended considerations of terraforming. Centripetal forces channel the early treatment of this motif along specific narrative trajectories, and as the tradition coheres it incorporates and generates other themes and plot elements, combining them with other sf tropes while further developing influences that have fed into early treatments of the motif. Hence, histories of colonialism linked to the theme of space colonisation frame the terraforming motif. These new Earths are often imagined as a garden, and draw on pastoral themes that have been re-voiced through the influence of the Gaia hypothesis on conceptualisations of planetary ecologies. As becomes apparent, the motif is heavily ramified and subject to centripetal and centrifugal forces that collapse multiple narratives while opening them up to considerable reformulation. The presentation of worlds as global chronotopes raises questions over the nature and efficacy of a global sense of place and is complemented by local and regional representations of spaces as constituents of this global chronotope. These nested spaces complicate the universalisation operative at the level of the global.

The political and ethical implications of science and technology are central to *Venus*. Terraforming, as the Mukhtars' dependence on the Habbers shows, involves the application of technologically sophisticated engineering principles, which suggest the dominance of a specific ethico-political orientation underlying conceptions of space. Technology (as applied science) is inseparable from value judgements and ethics; the technological adaptation of planets is tied to political factors that dictate the colonists' particular relations to space. This relationship changes as views towards and about science and technology change, resulting in a literature that, by the time of Sargent's and Robinson's trilogies, consciously reflects on the range of explorations of the social repercussions of science already pioneered by earlier works. This is a consequence of one of the formal features of the terraforming narrative: the intratextual juxtaposition of planets, and the range of intertextual spaces developed by the terraforming tradition, is mirrored by a range of politicised spaces and metaphorical worlds.

Domes on Venus: Chronotopes of Enclosure

As the adaptation of alien planets often requires some protection from hostile environments, domes have been a prominent feature of sf narratives of colonisation and terraforming. Gary K. Wolfe considers domes in the context of images of the confined city (1979). He suggests that

> Science fiction is hardly the genre we would look to for nature worship; and indeed there are numerous works in the genre that apparently condone such wholesale subjugations of nature as 'terraforming,' or the engineering of alien planets for the comfort of man. But even the expansionist motif of the genre can be seen as having its roots in an awareness of the finite resources of any limited environment, and the arbitrary growth of any institution at the expense of the natural world – as opposed to the purposeful expansion *into* the natural world – is viewed with scepticism. (Wolfe, 1979, 90)

Robinson's 1990s *Mars* trilogy and the popular success of Hollywood film *Avatar* (2009) may go some way to addressing this perceived absence of 'nature worship' in sf, but as has so far been shown, terraforming stories have not confined their enquiry to an ecological awareness of the finiteness of Earth's resources, but have explored the connections between wider socio-political and environmental philosophical themes. Nevertheless, Wolfe's suggestion that recognition of the finitude of Earth's resources motivates expansionism in sf offers a broader ground for environmental reflection on the connections between the ecosystem of artificial environments and natural, planetary ecosystems. The use of domes to explore the social dimension of terraforming draws on early representations of the city as dystopian, as in Wells's 'When the Sleeper Wakes' (1899) and Fritz Lang's *Metropolis* (1927). Yevgeny Zamyatin's *We* (1924) uses a dome-like contrivance – a glass city – as a literalised metaphor for the public scrutiny and regimentation by a repressive socialist government of all spheres of the individual's life. Such images of the domed city appear in Anderson's 'The Big Rain' and *The Snows of Ganymede*. The chronotope of the city often functions as an emblem for the ideological values of its nationalist government and a site for the clash between opposing visions of future social arrangements.

In *Venus of Shadows*, the historian Malik suggests that domes 'are a kind of experiment, since the terraforming of Venus could have gone on without them,' and on other occasions he refers to them as a type of social experiment (Sargent, 1989b, 107, 261). Although often imagined as an essential element of the terraforming project, their importance lies in certain resemblances between planets that are conceived as uncontained biospheres and the contained biosphere of the dome. Domes, as miniature worlds, shield the colonies from a planet that sets strict boundaries to terraforming and habitation. These limits metaphorically and literally correspond to political restrictions established by the Mukhtars: '[t]he people of the Project imagined a world free of Earth's constrictions, a world where, instead of displacing other forms of life, people would be creating life from lifelessness. But the domes would shut the settlers away from the dangerous world outside; the first Cytherians would be prisoners' (Sargent, 1989a, 403). These political limits are imagined as a form of displacement of life, while the terraforming project as a whole becomes a symbol for the renunciation of a political status quo confined to Earth. Domes, however, are ambivalently imagined both as spaces for developing a specifically Cytherian (Venusian) culture and community and as prison cities designed to curtail cultural and political deviation from Earth's governmental system. A similar ambivalence is present in *Mars*: at the end of *Red Mars* a Martian revolution for independence from Earth is suppressed by the strategic destruction of domes controlled by the revolutionaries, resulting in many casualties. The fragility of the domes and their status as containers that set physical, political and socio-cultural limitations on the colonists illustrate the correspondence between physical and metaphorical space while figuring an anxiety at the heart of civilisation's technological capacity to shape new worlds through an application of politically directed science.

Domes are a way to analyse the alien environment from a physically detached and separated 'objective' position. Yet they also shut that alien environment out, excluding it from the colonists' physically experienced sense of place (which is a good thing, considering Venus's fierce inhospitality). As protective structures, they narrow the colonists' imaginative horizons and turn the social experiment inward, towards the exploration of social crises. In *Venus of Shadows*, the domes become metaphorical crucibles where a repressive ecotheist cult rises to social dominance and suppresses the population. These crises arise as a consequence of Earth's attempt to explore the limits of the human: '[t]he domes seemed nothing more than a vast laboratory in which to test various subjects, to discover if people would willingly become prisoners of the dream of terraforming, to find out if human beings could transform other worlds without losing their ties to the old or shedding their humanity, as it appeared the Habbers might' (Sargent, 1989b, 185). The problem of developing a new culture on Venus is magnified when the otherness of the environment is excluded. This problem manifests as a failure of re-territorialisation, of developing a culture and identity that can allow the colonists to connect with their current local and global context while also providing a space for recognising their status within a system of interplanetary relationships. Instead, the experiment is directed towards Earth's continued control over the settlements, suggesting a refusal to respond in new ways to the deterritorialisation that colonisation entails.

Against the social pressures that develop on Venus, the appearance of life on its surface operates as an emblem for the limits of political control and for a re-territorialisation of Venus. In *Child of Venus*, several colonists notice something new and unexpected growing from the genetically engineered moss colonising Venus's surface outside the domes: 'it's life, and it's something we didn't plan for. We didn't plant it there ourselves, and none of our computer models, including the ones that allowed for possible contamination of our equipment, predicted that anything like it would grow from that moss' (Sargent, 2001a, 305). These computer models – themselves a form of world reduction that simplifies the complexities of ecological systems – open spaces that signal a new global context in which the scientifically informed awareness of the physical nature of Venus is emphasised. This is connected to the metaphorical implications of Iris's specialist subject, climatology, which offers a challenge to social stasis. For Iris, her early study of the terraforming project nurtures 'a sense of how much a tiny, seemingly insignificant event could alter a great deal; at the same time, it made her conscious of how much still lay outside humanity's control' (Sargent, 1989a, 89). This insight draws from the mathematics of chaos theory, a theme central to other ecologically focused terraforming stories such as Frank Herbert's *Dune* and Turner's *Genesis*. Chaos theory's fractal self-similarity at multiple scales is linked to terraforming via computer modelling, which uses chaos mathematics to simulate models of environments for testing. The emblem of the unexpected growth of plant life on Venus introduces a new landscape, one that underscores how the predictive capacities of scientific models fail to fully account for nature and its otherness.

This challenge to the colonists' expectations, to their view of nature, is accompanied by a literal challenge to social repression embodied by the Habbers, a genetically engineered, divergent and 'alien' human group who reject Earth's politics, government, and history, along with the planetary chauvinism of terraforming and colonising worlds. The terraforming project is an interface between two cultures that evolve unpredictably, mirroring the example of the unanticipated growth of plant life amongst the engineered moss outside the dome. The relationship between Earth and the Habbers is figured via a biological metaphor: '[e]ach culture was a cell, with Venus as the membrane through which molecules from each cell could pass. Without such an exchange, the cells would die' (Sargent, 1989b, 324). Some of the Habbers recognise the necessity of interchange between themselves and Earth, fearing they 'would lose that contact with a young, striving culture and become more insular [while] Cytherians, dreaming of their future world while trapped behind their domes, might turn on one another again; the cauldron could boil over once more' (514). The Habbers' unwillingness to completely break away from Earth and Venus suggests a lack accompanying their rejection of planetary nature and the type of communities that might develop in these spaces.

Wolfe's discussion of artificial worlds as one of sf's important devices can be usefully compared to Sargent's use of the dome. In the *Venus* trilogy, connections are made between the 'Island' space stations in Venus's atmosphere, the later development of domes on Venus's surface, the space station Anwara – designed and built by Earth's technicians – and the Habbers' artificial worlds. The dome's surface can be read in terms of Wolfe's icon of the barrier, which shields the colonists from the unknown of the Venusian environment. Wolfe's claim that 'in many stories the barriers are simply temporary blocks to humanity's inevitable conquest of the cosmos' is certainly relevant here, as the Islands and domes are intermediary spaces leading towards unrestricted habitation of the planet's surface (1979, 34). The Islands, designed to settle on Venus's surface as new domed settlements at certain stages of the project, embody this idea of a series of barriers.

There are also important differences between these devices, however: Mahala in *Child of Venus* 'felt far more vulnerable here [on Anwara] than behind the transparent dome of a settlement or an Island,' a feeling she attributes to a complex of factors including, controversially, 'more faith in the Habber technology that had created the ceramic-metallic alloy of the dome material' of the Islands and settlements (Sargent, 2001a, 211). The Habbers' mobile artificial worlds, in contrast to the rooted domes, overcome other significant barriers, having historically carried their inhabitants away from a devastated Earth and into space. There is an ethical component to their refusal to inhabit planets, expressed by the Habbers' belief that '[t]o make use of some planetary resources was acceptable; to alter a world completely was unnecessary and undesirable.' They are a society on the cusp of crossing a barrier that would definitively separate them from their planet-bound roots, and yet they invest in terraforming Venus: '[t]he Project provided the Habbers with their only direct contact between themselves and the people of Earth; it was their last link with the rest of humanity' (Sargent, 1989b, 37).

Part three of Robinson's *Red Mars* ('The Crucible') and six of *Venus of Shadows* ('The Cauldron') use metaphors of a heated vessel to signal the transformative effect of the device of the dome on its contained community, foregrounding themes of social conflict, experimentation and integration. The problems that arise when working towards this potential synthesis form the most significant barrier in Sargent's and Robinson's terraforming narratives, and it is the potential for this social synthesis that encourages the Habbers' continued investment in Earth's grand engineering endeavour. The dome provides a space to explore the goal of forming new cultures from a synthesis of multiple groups with different and often opposed agendas; this chronotope is extended to the planet as a whole. In *Venus of Shadows*, one of the children born on Venus tells of his father's description of Earth's sky as 'sometimes like this huge kettle over the Earth,' to which his friend Teo replies, '[t]hat makes it sound like a dome' (Sargent, 1989b, 306). The relationship between planets and domes is echoed more frequently in *Blue Mars*, where the relatively gentle Martian climate allow the colonists to walk upon the planet's surface. To a greater degree than the domes in *Venus*, which completely cut the colonists off from a physical relationship to their environment, Mars in the *Mars* trilogy, with its clouds 'defining the dark dome of sky above them,' incorporates nature and a plurality of landscapes within its global conception of space (Robinson, 1996a, 193).

Robinson uses the physical relationship between the planet and the colonists to engage in a sophisticated enviro-ethical critique. Nevertheless, even if specific environmental ethical discourse plays little part in texts prior to this period, environmental enquiry has been a longstanding feature of many terraforming narratives. The emphasis Robinson places on the landscape as an environment that provides some of the colonists with a deep personal connection to Mars helps illuminate one dimension of the Habbers' investment in terraforming. Drawing on pastoral chronotopes of the garden, the Habbers' space habitat is described as 'a world of wide corridors, simple rooms, and a garden of forests, lakeshores, hills and plains at its center that seemed meant to be a monument to Earth' (Sargent, 2001a, 39). Their artificial reproduction of a contained biosphere can then be read as an attempt to retain a connection to 'nature' and to Earth's planetary spaces, since such biospheres need not replicate Earth's ecology. The Habbers' reluctance to sever themselves entirely from planet-bound nature can be seen as a response to a deterritorialisation that threatens to disconnect the Habbers from an interplanetary sense of community.

The distinctiveness of the terraforming motif is to some degree founded on its propensity to create spaces where the confluence of the three domains of politics, science and environmental enquiry is explored in the context of world-building. Terraforming is a metaphor for the practice of building sf worlds as much as a textual space for experiments in engineering societies. Nirgal, as the first Mars-born who visits Earth, says in a speech in Robinson's *Blue Mars* that 'we have to help each other. We have to regulate ourselves, we have to take care of the land. And it's here, in this part of the project, that Mars can help Earth. First, we are an experiment in taking care of the land. Everyone learns from that, and some lessons can be applied here' (Robinson, 1996a, 197). By experimenting with new relations on Mars and by making use of insights that are derived from environmentally inflected experiences of new social and physical relations to others and to the land, the *Mars* and *Venus* trilogies use terraforming to explore barriers to habitation – the making of homes in a universe that humanity is alienated from. The habitats in the *Venus* trilogy, chronotopes that symbolise escape from Earth's restraints, are metonymies for a culture that has alienated itself from environments that offer spaces to experiment and create new ways of 'taking care of the land' and of each other. Terraforming is distinctive, then, for its synthesising tendency; it is a motif that gives play to an abundance of social and environmental parameters for transformation.

The Pastoral in Pamela Sargent's *Venus* Trilogy

The *Venus* trilogy uses a mixture of Bakhtin's pure types of idyll and employs the chronotope of the farm and garden to characterise the terraforming of Venus. Images of both have been used in earlier works to embed value into their representations of spatial adaptation. Speaking first of the chronotope of the garden, *Venus of Dreams* uses the practice of science as one element of the craft-work idyll. Sargent combines Carl Sagan's 1973 proposal for using algae to seed life on Venus with a pastoral garden image: '[l]ife was in the dark clouds, where the strands of algae still fed on the poisons. There, humanity's microscopic garden had taken root in the stormy atmosphere' (Sargent, 1989a, 366). This ecopoietic image promises a burgeoning onto other spatial levels of the garden chronotope and is paralleled by a cultural mirror of the garden that draws on metaphorical applications of the scientific notion of 'the built-in amplification of biological replication' (Sagan, 1973, 513). This 'amplification' can be metaphorically applied to the domes, which mirror Earth's social relations and function as experimental spaces for the development of new communities. Domes, by metaphorically turning the colonists' gaze inward, amplify Earth's moral landscapes and social values. The Habbers' gardens, and the gardens of the domes and the Islands, differ in a number of ways, but the chronotope of the garden within the artificial world points to a fundamental anxiety rooted in alienation from the environment and an amplification of social conflict.

In *Child of Venus*, the narrative's focus on terraforming is supplanted by the discovery of an alien signal from deep space. Earth and the Associated Habitats agree to construct the 'Seeker,' a space habitat designed to carry a joint crew of Cytherians, Earthfolk and Habbers to the signal to establish first contact. Because of the distance of this journey, the crew will ultimately be isolated from their familiar time by over a millennium, leading the Cytherians and Earthfolk to accept life-extending treatments from the Habbers in order to survive the mission. Their return to Earth's solar system represents a vast period of temporal deterritorialisation. This myth of sf, of first contact between alien civilisations, is deflated when the signal is discovered in a lifeless solar system and, falling silent, affirms the banal: that 'humankind knew of this alien intelligence without being able to reach out to it' (Sargent, 2001a, 418).

This variant of the space exploration theme embeds another experimental space within the terraforming narrative to explore the overcoming of boundaries between the Cytherians, Earthfolk and Habbers. The failure of contact is a deeply ironic plot development that underscores the importance of making homes in contradistinction to the exploration of deep space. The structural advantage of this plot development for representations of terraforming lies in its compression of the project's timescale, thus allowing the explorers to return to a fully transformed Venus and to experience it from their now alien perspective. The apprehension that the explorers feel regarding the now unknown conditions of the human inhabited solar system leads the protagonist Mahala to think that 'they had even more of an obligation to build a true human community aboard the Seeker' (Sargent, 2001a, 425). This realignment towards constructing a 'true community' takes place in a pastoral space: '[t]he core of the Seeker was made into one of the gardened environments so beloved of the Habitat-dwellers, a very gently curved landscape of rivers and forests and open grassy land' (397). Although this artificial world mirrors that of the domes and of Venus and Earth, it differs in the degree of its separation from humankind's history and continuing influence. The six-hundred-light-year distance between the solar system and the alien signal imbues the chronotope of the Seeker with a temporal and spatial isolation that contrasts with the domes, which are connected to others in a network of communities. This space offers a pastoral separation from the complexities of globalised politics on Earth and Venus. However, the Habbers and colonists depend on a connection to planetary space in order to develop their sense of being at home:

> such places embued them with a feeling that they had not lost their humanity and their natural past entirely, while for the Earth-people aboard the Seeker, the landscape would function as a reminder of home. Mahala, who had lived her life in the environments of Venus's domed settlements, thought of the core as a monumental and ecologically complex version of a garden or a park. (Sargent, 2001a, 398)

The Seeker is a landscape subject to polyphonic interpretation and a space that consciously symbolises the sedimented voices of culture and their interpretations of nature.

The chronotope of the Seeker exemplifies another use of the icon of the Habitat to reflect on what Bakhtin identifies as the first distinctive trait of the pastoral: the portrayal of a community and their connection to their environment. In contrast to the cultivated landscapes of the contained biosphere of the Seeker, the wild nature of the terraformed Venus that they return to highlights the sense that the pastoral spaces of the domes and habitats are somehow insufficient for supporting the colonists' growing attachment to their environment and community. 'Home,' the final section of the trilogy, connects the pastoral to the theme of making a home, but continues to embed the idyll within the sf framework to offer an alienated perspective on the familiar nature of the solar system. Mahala is aware that '[s]he and the others with the Seeker were outside of their history, and it might now be impossible for them to rejoin it' (Sargent, 2001a, 431). The sense of a coherent community identity is thus in part a function of a group's participation in history, a notion that Allaby and Lovelock touch on in the concluding chapters of *The Greening of Mars*. Memory becomes central to building this sense of community; terraforming in this context serves to obliterate this sense of history. Echoing Allaby and Lovelock's discussion of speciation, the returning explorers encounter humans so fundamentally divorced from what they consider human as to constitute a different species. The history of these posthumans is sealed behind an impenetrable temporal barrier, and any possibility of communication leads only to ambiguity. Once again, interchange between two alien cultures is denied, although the posthumans are content to allow the explorers to settle on the terraformed yet mysteriously uninhabited Venus.

The portrayal of communities in the *Venus* trilogy is tied to Bakhtin's family idyll through the texts' focus on new generations. Mahala links terraforming to this pastoral theme when reflecting that '[t]he instincts that had given [Mahala her] daughter, that had given that genetic tie to the past, had also given birth to this world' (Sargent, 2001a, 445). The theme of new generations that concludes each novel connects the family idyll to terraforming through a shared temporal orientation rooted in anticipation of the future. The long-term terraforming of Venus is a project that, at its ideal, is directed towards the future and humankind's descendants, a gift that 'so many of [the explorers'] ancestors had labored to create' (445). Mahala's child Angharad, named after her ancestor in *Venus of Dreams*, gestates ectogenically during the return home and belongs to this new generation whose separation from Earth's history makes them fundamentally different to the other explorers. Her experience of a now deserted yet habitable Venus, spotted by the remains of the domes and the single memorial pillar commemorating her ancestor Iris, is preceded not by fear or apprehension, as in her parents' case, but by 'wonder and curiosity' (430). History begins again; it is as if the utopian promise could not be fulfilled without a journey that would join the Cytherians, Earthfolk and Habbers in a space divorced from the corrupting influence of the wider interplanetary community. Physical terraforming parallels social terraforming, which is made possible through two other grand social endeavours, deep space travel in artificial worlds and the discovery of alien life.

While the narrative chooses to diverge from portrayals of terraforming and the act of building new communities on Venus, *Child of Venus* uses a deflationary strategy towards common sf tropes to help underscore the wonder with which the new generation approaches Venus and the construction of a planet-bound community. Recapitulating another prevalent trope of sf, Benzi, Iris' Venus-born son who had abandoned the planet to join the Habbers, reflects on the consolatory urge to parenthood as an expression of larger biological workings: '[l]ife always finds a way [...]. That was certainly one of the assumptions behind the terraforming of Venus, wasn't it? Give life an opening, a chance, and...' (Sargent, 2001a, 428). Although the trilogy ultimately turns away from the process of world-building as a community activity, it ends with the explorers ready to take up the project with the knowledge that other alternatives lead back to the community and their environment.

Venus of Dreams begins in Lincoln, the farming village where Iris's mother, Angharad, plays an important role as representative of the village's values and traditions. As one of many Plains villages, this community provides food for Earth's population. The Plainspeople take pride in this role: Angharad explains that 'we're just about the most important people in the world, aren't we? We feed most of it, and it's our tongue people of different lands use to speak to each other' (Sargent, 1989a, 26). The centrality of the farms for the global civilisation is linked to the widespread use of the Plainspeople's language and contextualises the shape of this society's history. A discontinuity exists between the image of a space contemporaneous to the reader and the events of the trilogy, a historical period where resources were fought for between nations and where the language Angharad refers to, 'Anglaic,' developed wide usage before the New Islamic States eventually emerged and secured worldwide dominance. Anglaic has left its traces, along with its political implications, upon the text's cultural topography, where a constructed historical continuity to a time contemporary to the reader can be inferred. The traces of this socio-political history is important for terraforming narratives and for the agricultural pastoral, as it impacts upon the motive and social function for engaging in terraforming. The farm, seen as a space for the production of resources, is represented within a frame that emphasises both the metaphorical and the literal transformation of culture over time.

The Plainswomen connect agricultural space to generational time by conjoining their farming culture to an unbroken matriarchal line. The importance of seasonal cycles and the labour that this entails, along with cycles of generational time, subordinates a linear time represented by a matriarchy in which daughters assume their heritage when their family farm is passed to them. This ties the family to the landscape in a continuing cycle that is threatened by Iris's wanderlust, which she directs towards Venus. Hassler argues that while the Habbers embody a utopian future, notions of lineage and heritage are 'complemented by a sense of mission back to the fragile Earth environment. What the settlers learn from terraforming Venus can be used to avert further ecologic damage on Earth' (1997, 154).

Iris's wanderlust disrupts this attachment to the local, recapitulating the pastoral's traditional motif of the idyll disrupted by technological intrusion, but it also introduces a new conception of a global, interplanetary space that relates uneasily to the Plainspeople's sense of inherited place. As a child Iris cannot conceive of departing from the farm and the Plains. She reacts conservatively when her grandmother Julia reflects on Venus and imagines a static future where '[t]he boys would wander the Plains and the girls would stay in Lincoln and farm, as her family had always done' (Sargent, 1989a, 4). Angharad, like the other Plains families, is acutely conscious of the continuity of her ancestry, which is 'preserved not only in Angharad's memory but also in the memory banks of the cyberminds that served Lincoln' (4). Sf's use of the pastoral, even when the country is represented, is interpenetrated with items associated with the city: 'cyberminds' are the artificial intelligences that operate as the Nomarchies' information management technology, while the farm's role in supplying the global population with food has already been considered.

Angharad's failure to prevent Iris from leaving Lincoln to pursue her scientific studies demonstrates this vital link between the space of the farm and the generational cycle. The link between generations and the local landscape comes into conflict with Iris's awareness of a space outside the local. The idyll of the agricultural pastoral is confronted with the growth of a new conception of space that implies freedom, large-scale achievement and grandeur, all of which are accompanied, in Angharad's view, by 'the misery and regret that were sure to come to [Iris] if she did not do her part for the farm and her line' (Sargent, 1989a, 154). Iris confronts her mother with an alternative vision that accounts for her disruption of this cycle, arguing that '[a] grandson of yours might take a branch of our family line to another world' (155). The divergence of her family line and the pride that Iris's achievements bring to her Earthbound family redounds on her descendants: Harriett Teresas informs her cousin Mahala (Iris's great-great-granddaughter) of the greater worth that knowledge unrelated to agriculture is accorded by their Earthbound family.

The process of reconciling the divided family branches occurs in a section called 'The Garden,' thus establishing a series of links between the two central pastoral chronotopes of the garden and farm and the generational, family idyll. Joined to the successes of spatial adaptation that the terraforming project has wrought, changes of behaviour for those on Earth have also reverberated from the decisions and actions of those involved in the project. This cultural change arises in the interstitial space created by the tension between time as cyclical and stable and time as linear and subject to historical change, thus showing how the pastoral reconciles these apparently contradictory temporal views: 'Teresa also encouraged my lessons because she saw that a time might come when I'd have to know more than how to run a farm. "Things can change," she always said. "They've been the same for a long time now, maybe for too long." And as it turned out, she was right' (Sargent, 2001a, 297). This response to the evidence of inevitable change to Earth's global society does not end by retreating into the idyll of the idealised, sublimated pastoral, but leads to a flexibility to adapt to new social conditions initiated from a space beyond the local. The stasis fundamental to the survival of the global population after the Resource Wars is now a restraining social enclosure. By following Iris's lead as the first pioneer of change in the family, the succeeding generations respond to the demands brought about by shifting societal needs at the global level, both internally and in connection with the repercussions of the terraforming project.

Iris links terraforming to farming and the garden when reflecting that '[m]aking grain grow on the Plains was little compared to seeing a world bloom under one's hands' (Sargent, 1989a, 13). Again, the desire for a magnification of achievement is sustained through an implicit depreciation of the local on Earth. The verb 'making' emphasises the work involved in farming on Earth, whereas the verb 'seeing' foregrounds a passive perceptual act that downplays the labour involved in terraforming Venus, thus idealising the project. The nature of the activity also shifts from agriculture, a resource-focused form of cultivation, to gardening, a form of cultivation associated in functional terms with leisure. This metaphor captures the shift of value from the sf use of the agricultural pastoral to the garden pastoral. Earth is envisaged as a space requiring constant resource management. The scarcity of resources provides one of the most powerful motives for terraforming other planets; visions of the garden pastoral on Venus elide the stark economic challenges involved in terraformation.

When Iris has matured and is living on the Islands she thinks of the connection between farming and terraforming explicitly: '[t]erraforming, in its own way, was an extension of farming, a way of forcing a world to yield to the needs of human beings. Venus, like Earth, would surrender to their efforts for a time, and then strike back at them; each season would bring its own battles' (Sargent, 1989a, 257). Terraforming for Iris is imagined not as farming in a metaphorical sense, but as an extension of the history of agriculture and physical landscaping. This arises as a logical consequence of civilisation's technological ability to manipulate its environment and the accompanying view that human relationships to the land involve active domination. Venus, like Earth with its unpredictable climatic changes, is not simply passive but is imagined as responding in cycles of aggression to this attempt at domination. The chronotopes of Venus and Earth are, when seen in their totality, personified. A scientific understanding of climatology and a theory of ecology that emphasises the relationships between complex systems are incorporated into this vision of terraforming: '[s]he sighed. She had a model to study, one that showed the cycle of photochemical and thermochemical reactions that changed Venus's sulfuric gases into cloud particles. The cycle was complex, and a recent increase in volcanic activity on the Venusian surface meant that more sulfur, emitted by volcanic eruptions, would be entering the atmosphere' (Sargent, 1989a, 234). This model encodes ecological principles of interconnectedness that mirror the economic integration of Lincoln in a network of villages, all of which are functionally directed towards the political structure of the Nomarchies: '[t]here had not been as much of a surplus that year, but the weather in other parts of the world had been favorable; the Nomarchies would be able to feed all of Earth's citizens' (Sargent, 1989a, 48). This view of the Plains differs from Angharad's in that Lincoln's dependency upon the larger patterns of climate is underscored alongside the Nomarchies' dependency upon both the Plains villages and Earth's planetary ecology.

These relationships define another spatial reordering of the divided landscape in terms of a layering of orders of magnitude and function. The Nomarchies are organised as a network of politico-cultural spaces (such as the Native American Nomarchy) encompassing the physical spaces represented in the text (the local, regional, climatic and global). The interdependency of local space, of the different farms and villages, and their relationship to the regional level, is given an sf twist: '[a] few rainclouds might be routed their way if the rain were not needed more elsewhere and if the task could be done without altering climatic patterns too greatly. The Nomarchies tended to be cautious about such matters' (Sargent, 1989a, 40). This portrayal of the global management of Earth's climate and weather contrasts with Norman Spinrad's depiction in *Greenhouse Summer* (2013 [1999]) of an Earth where incompatible national and group interests make such synchronisation impossible to manage. Sargent's use of geoengineering develops the relationships between the complex, chaos-informed understandings of climate and of ecology, in both their scientific and their metaphorical senses, via an emphasis on the economic and political links between spaces. Evolution plays its part in this dialogue. As mentioned above, terraforming is conceptualised as an advanced form of agriculture. The term 'planetary evolution' appears in *Venus of Dreams* four times and refers to planetary-scale climatic, geological and ecological cycles and their interconnections (Sargent, 1989a, 59, 195, 235). Terraforming is a form of long-term labour in which a dynamic conception of time is essential to the transformation of these landscapes. The *Venus* trilogy exemplifies sf's use of the garden and farm chronotope to bring into dialogising relationships new scientific discourses and traditional uses of the pastoral.

Frederick Turner's *Genesis: An Epic Poem*

> Their acts, mortal and cast away, / Are crystalled in the melt of history (Turner, 1988)

Frederick Turner's *A Double Shadow* (1978) is a terraforming novel set on a Mars whose inhabitants, having succeeded in conquering space, have turned to a polyphony of aesthetic principles to structure their lives and to continue to extend the limits of their powers. Turner also wrote *The New World* (1985), an epic poem set in a future America that has been divided into independent county-states based on Jeffersonian principles. This section, however, focuses on Turner's epic poem of terraforming, *Genesis*, which was influential amongst researchers at NASA's Houston Space Centre when first published. This popularity eventually led to his contribution to the founding of the Mars Society and an invitation to the 1991 terraforming workshop at NASA's Ames Research Centre.

Genesis tells the story of the terraforming and colonisation of Mars, and of the ensuing conflict between Earth and its interplanetary rival. Unusually for sf, the narrative is recounted in the form of an epic poem, whose ten thousand lines consciously align this futuristic epic to the Classical tradition of epic poetry. Other Greco-Roman influences are significant: Lovelock's Gaia hypothesis, with its implicit mystical element, alludes to the Greek goddess and helps furnish the narrative with an ecotheist movement that provides an ideological opposition to the conduct of science on Earth and amongst the Martian colonists. The tradition of the bard inspired by a Muse to recount the stories of heroes is transferred into the future: two narrative voices, one of a narrator located in the future of the events told of in *Genesis* and one of a narrator contemporaneous with the writing of the poem, resonate with the cosmological view of space and time adopted by the Stapledonian narrator of *Last and First Men* (1966 [1930]). Turner expands the notion of the classics to include other cultural forms of epic, including 'the Uruk of the composition of Gilgamesh, the Israel of David and Solomon, the North India of the Mahabharata, the Japan of the shoguns, the high Mayan civilization, Augustan Rome, High Medieval Nordic Christendom, Elizabethan/Jacobean Britain, and Nineteenth-Twentieth Century USA' (see Pak, 2014b, 7). He explains that these periods

> were moments when human civilization, in its cultural, political, military and economic expansion, had the confidence to look at its own flaws and dark origins, the hope to envision both ancient and future expressions of its essence, the leisure to dream, the

> philosophical largeness of debate to think about the whole human race and its universe, and the exuberance to celebrate the glories of nature and culture. (See Pak, 2014b, 7)

These epic tropes are complemented by reflections on the conflict between space colonisation seen as an extension of westward colonial expansion across the American continent and space colonisation seen as an extension of an organism's tendency to expand outward from a point of biological origin. The polyphony of cultural traditions feed into the landscaping that occurs on Mars, illustrating how terraforming works to bring together a range of voices that provide a basis for establishing a sense of home by re-territorialising the planet.

Patrick Parrinder argues that sf is a form of 'truncated epic' that does not often inherit the epic's amplitude, but is concerned with 'future or alternate history' and the fate of 'whole societies or of the human race, its collaterals or descendants' (1980, 93). Vandana Singh draws on the Hindu epic *Mahabharata* to suggest that 'climate change also calls for epic stories, spanning generations and ages, complete with melodramas of the human and natural variety'; in the same roundtable discussion, Robinson agrees that 'thinking of [sf] as a vast epic is better' than his own metaphor of sf as an encyclopaedic 'wiki' (see Harrison, 2012). Adeline Johns-Putra argues that sf film grows out of a tradition of Classical and biblical epic film (2006, 208–213), while classicist Judith de Luce reflects that 'I said that terraforming was material for a movie, but it is even more natural material for an epic' (1993, 20). In his 2012 study *Epic: Form, Content, and History,* Turner himself points to Arthur C. Clarke's *Childhood's End,* Isaac Asimov's *Foundation* and Kim Stanley Robinson's *Mars* trilogy to argue that '[t]he science fiction genre is lavishly epic in scope, inspiration, and action, taking on the ancient themes of world-creation, sacrificial heroism, death and immortality' (16).

Turner reflects on developments in cultural and political theory that have resulted in the dearth of critical attention paid to the epic. They have culminated in the rejection of grand narratives by postmodern critics, who consign epic to what Jean Baudrillard, in *The Illusion of the End* (1994), has called 'the dustbins of history' (Turner, 2012, 3). Turner argues that if the planet is as Baudrillard suggests – 'a dustbin of narratives' – then that is where humankind must make its home, and the struggle to do so 'might be a fairly grand story in itself' (3). Narrative viewed as a Foucauldian archive, or what Thierry Bardini theorises as 'junk' (2011), presents a resource of elements potentially viable for transformation into new forms for the contemporary age. In Turner's words, 'the past is the source of progress, an archive of possible strategies and narratives and collective memories to be consulted and recombined to meet the new situation' (2012, 102). Epic is a fundamental feature of that past, a highly conscious artistic creation that Turner considers 'the most fundamental and important of all literary forms' (13).

Turner's view of narrative as a way to develop modes of habitation on Earth supports de Luce's claim that the motif of terraforming is particularly suited to epic poetry, concerned as it is with the theme of making a home on other planets. Turner writes that the epic journey, involving exile and a 'nostalgic yearning for home,' invokes some variation of a fall and makes of the epic a search for a means whereby humanity can create new homes on alien landscapes (2012, 169). For Turner, the construction of a self-referential and self-validating poetic identity over the course of an epic poem makes of it 'a gigantic form of lyric' that is 'far more "dialogical" than any novel, since its Gödelian questioning of its own premises leaves it utterly free for anyone's possession' (39). Turner's resistance to postmodern theory lies in what he sees as its failure to recognise the openness of tradition, leading it to de-value prior cultural forms. He argues that '[e]pic is always already post-modern [...], but it has not had to concede any of its veridical power in the process' (39).

Turner departs from traditional formal and compositional definitions of the epic, preferring instead to focus on content, subject, and meaning. He detaches the notion of the Jungian collective unconscious from Joseph Campbell's monomyth, instead substituting as the basis for the monomyth 'the more provable unity of our common human evolutionary history' (2012, 11). Turner's emphasis on the steadily accumulating body of scientific knowledge in this field attempts to address the risks involved in appealing to evolutionary history. Responding to what he calls 'the shadow that has fallen over myth studies,' Turner builds on the structural anthropology of Claude Lévi-Strauss and on studies of 'the source and enactment of story in ritual' pioneered by his own father, the cultural anthropologist Victor W. Turner, to lend weight to his assertion that epic is 'the history in symbol and story of the human species itself' (12). Turner's statement of his view of the meaning of epic is compelling subject matter for the composition and criticism of sf epics: 'epic is basically about human evolution – that is, epic is the traditional way we have explained to ourselves as a species our emergence from nature and the stresses within our own nature that result from that emergence and our look back at it' (Turner, 2012, 8). Turner sees epic as one of the chief arts that worked to establish a feedback between evolution and culture, helping humanity to modify the quality of its adaptation from non-human origins. Turner's conviction that this cultural practice helped shape humankind finds expression in the creative work of the

Martian colonists in act V, scene ii of *Genesis*, 'Evolution and the City,' where the narrator describes how 'They found the arts that made the cavemen human / And strained our chromosomes to genius' (Turner, 1988). The epic is one of these arts, a highly prized form to the Martian colonists, who have immortalised the events told of in *Genesis* in a fresco displaying the narrative's scenes to Martian immigrants. This view of epic resonates with definitions of sf from critics such as Brian Aldiss, who emphasise sf's attempt as environmental fiction to tell the story of humankind's relationship to the external world (2001, 4).

The birth of the Sibyl and description of her role as prophet for the Martian community concludes the epic action of *Genesis*. The narrator's description of the early stages of Martian colonisation chime with Turner's own view of the disproportionate value attached to postmodernist theory in contrast to the epic: 'Mars had its brief, silly postmodernism. / Then as the Sibyl spoke the cosmos blazed / With its mysterious, bounded clarity' (Turner, 1988). This origin story aligns artistic creation to the creation of the new Martian republic and, at a greater scale, to the creation of the cosmos itself: a recreation that redefines the boundaries of the conceivable by realigning the human relationship to the boundlessness and mystery of nature. Beyond this new 'bounded clarity' lie further mysteries gestured towards by the openness of this rebirth, by the recognition that the future, 'Its mutual, guessed, but still intelligible / Working out of its own destiny,' admits the possibility of further societal permutations and philosophical insights that would redound on civilisation, goading it, in Stapledonian fashion, 'To dream the higher dream of consciousness':

> And as the artists worked they found the genres
> Rooted in the grey loam of the brain,
> Where flows of value knot to branch and bole,
> And differance can make a difference,
> And flowering risks a nest egg of tradition; (Turner, 1988)

The epic is arboreal and organic and is mirrored by the structure of the human brain. Patterns of neurons provide the infrastructure for 'flows of value' which, in 'the grey loam of the brain,' are composted and branch into numberless possibilities, a Derridean 'differance [that] can make a difference' to human culture. In line with postmodernism, Turner rejects the goal of metaphysical certainty, preferring instead the openness of evolutionary branching and cultural polyphony. Such deferral of metaphysical signification does not, however, invalidate the value of culture; it is instead made valuable by its embeddedness in a landscape of tradition that, through the actions of the epic hero, collides with what could not be or had not been incorporated into the cultural horizons of a given civilisation. If *Genesis* explores one future among many as a way of allowing us to map a more desirable route as an ideal to work towards, the self-validation of history that is so important for the epic poet would appear to be abrogated. In this sense the poet sacrifices his future for the possibility of another, one in which he will not exist. This sacrifice is a kind of self-validation of the poet equal to the more traditional kind.

For Turner, the epic is an act of sacrifice; the hero's sacrifice and the work of the epic is driven by the notion of commutation, which he explains 'has much in common with the processes of metaphorization, symbolization, even reference or meaning itself' (2012, 217). The commutation of sacrifice allows traditional meaning to be converted or translated into terms that can be understood within the bounds of a culture's signification. Feedback between the known and unknown, a concept resonating with Wolfe's discussion of the fundamental dynamic at the heart of sf's narrative mode and epistemological foundation, is therefore central to the work of the epic, which situates itself within the liminal, tension-filled space where meaning is constantly in flux. The work of '[t]he epic melody and sacrifice' is this periodic re-valuation of meaning, a reworking of tradition in the light of the epic protest: the acknowledgement of the boundedness of culture and the existence of resources that would defer the closure of a civilisation's cultural horizons. Turner explains that the *chijikijilu* of the Ndembu, 'the term for a fundamental religious symbol (what Catholics would call a sacrament),' literally refers to a blaze cut into a tree as a signpost to mark the incorporation of the unknown landscape into the known (see Pak, 2014b, 6). '[I]f the known world, the speakable world we have language for, is the past, and the unknown and unspeakable is the future, the blaze is in a sense constitutive of the present moment, to the extent that the present moment is that which mediates between the past and future' (7). In this context, the terraforming narrative is just such a blaze, a way of marking the unknown and incorporating it into the imaginative horizons of a culture. It is a form of landscaping that embeds cultural value into the environment and works to re-territorialise the unknown.

Poetic description of the calderas of Greece and Italy, fecund with the landscaping of mythological and epic tradition, is connected directly to the colonisation of Mars through a tradition passed from the Greeks and Romans, through Europe and onto America and beyond. After eleven lines that invoke specific features of the Campi Flegrei caldera in Italy, the poet announces that 'These were a Mars to the exploring Greeks, / Who set their cities on a seaward hill, [...] / And sowed the place with myths and oracles' (Turner, 1988). Thus, Greco-Roman landscapes are made to answer to a new context on Mars. History – memory and experience recorded in myth and epic – becomes an archive of possible structures and interactions with the Martian landscape. This image of cultural succession raises the spectre of American 'manifest destiny,' which maps the dynamics of a series of localised cultural developments onto the total history of human evolution. Turner's evolutionary perspective, his extension of the term 'Classical' to include other cultural traditions and his discussion of American colonisation in 'The Humble Bee: Restoration as Natural Reproduction' and elsewhere make clear his awareness of the risks associated with drawing such connections. Rather than the colonial justification of oppression that ideas of manifest destiny have historically supported, *Genesis* invokes this parallel to raise the idea that humanity is poised between exploratory and homemaking tendencies that pose a fundamental ecological dilemma. Turner points out that '[h]uman art, human fiction, human invention, human technology, are not unnatural forces that have suddenly erupted into nature, but are the natural continuation of nature's own evolutionary process' (1987, 16). The value of natural landscapes and the problem of 'naturalness,' given the long history of the human adaptation of the environment, connect the ethics of terraforming to the longstanding project of physical and intellectual landscaping.

Description of the gardens that Beatrice creates on Mars emphasises the polyphony of landscapes that Robinson later develops in his *Mars* trilogy. They are 'variations of Arcadia' that are temporally truncated, 'Branch[ing] out like palm-sprays from the parent stem' (Turner, 1988). Cast in a variety of modes, including 'The classical, heroic, and grotesque, / The meditative and the modernist,' these gardens take root in a Martian landscape already resonant with the landscape of Greece, Mars's regions having been predominantly named after such locations as Arcadia itself (according to the schema of the United States Geological Survey, which divides the Martian surface into thirty quadrangles) (Turner, 1988). Nevertheless, Beatrice deploys other cultural gardening techniques: 'the Islamic gardens of the Taj / And those Khmer quadrangles of tropic flowers, / Those stone arcades of tantric statuary / Are pressed for metaphor and simile' (Turner, 1988). These allusions to Islamic, Cambodian and Hindu gardening practices and philosophies are later complemented by Buddhist and Pacific Asian cultural practices when the narrator describes 'landscapes made for meditation / [...] where the pines / Bonsai'ed to knotted dwarfs [...] And its [the mind's] each stroke of thought's as natural / As leaves a bamboo brushstroke leaves behind' (Turner, 1988). These gardens, like the Pacific Asian art of calligraphy, write into the Martian planet human cultural landscapes that invoke a wide range of traditions. Modern imagination of Mars is not excluded; these traditional practices are joined by the construction of 'Canals, in honor of the ancient fiction' (Turner, 1988). This jumble of landscape features reflect on one another and exemplify how the colonists spatialise their reflection on history itself, an outcome of their sifting of those valuable parts of tradition to be incorporated into the new Martian culture. The net effect of these superimpositions of landscape is one in which 'All is in classical humane proportion' (Turner, 1988).

Beatrice reserves the 'heroic mode' of landscaping for the grand, sublime features of Mars, the Valles Marineris, the 'three grand calderas' of the Tharsis Ridge, Olympus Mons, and the new seacoasts of the terraformed planet. She treats these with 'the puritan iconoclasm / Of the American parks and wildernesses,' sometimes etching figures into the landscape 'like a Nazca god / Or neolithic Wessex henge-maker' (Turner, 1988). This sense of iconoclasm in the face of the sublime is complemented by the peculiar effects of the Martian environment on life itself, where 'The very giantism of the trees / Released from Earthly gravity, can give / The comic scale to landscapes of the sun' (Turner, 1988). The heroic and comic are balanced at this late stage of the terraformation of Mars in contrast to the tragedy of the Earth–Mars conflict preceding full independence. The towns that arise on Mars are arboreal in nature, being built into these alien trees 'Like Adirondack lodges years ago,' yet the poet notes that the public buildings, like the southern African architectural traditions of Great Zimbabwe, are constructed of stone: 'Their classical facades and fitted ashlars / Meet for the canons of the new republic' (Turner, 1988).

Although the term 'classical' in this context points towards strong Greco-Roman influences on the portrayal of a future Mars, its unobtrusive lower-case type and the appearance of classical elements from other cultures shift the designation of the term to a more abstract category that refers generally to those highly developed periods of any civilisation. The terraforming of Mars, it is implied, is the event that makes possible a futuristic classical age better able to embrace this wider range of cultural traditions. Alongside these traditional influences from the history of human culture are the 'Promethean disfigurements' of terraforming ('planoforming') and robotic mines, which the narrator somewhat unusually claims 'Can give the crowning shiver to a valley / Cragged with the terror of the shadow of death' (Turner, 1988). It is significant that these overt technological presences are aligned with death and the underworld, a location that in Turner's view is essential to the epic. It is from the underworld that the epic hero ascends and returns to the world of the living, bringing to civilisation new knowledge from the unknown. Despite these human interventions on the Martian landscape, 'much is left to chance and to the weather,' to the twin forces of (computerised) biological technology and ecology: 'Ganesh's biotech is everywhere, / Balanced by Charlie's crisp ecology' (Turner, 1988).

De Luce has argued that Turner's reliance on Arcadia as a model for Mars is problematic because 'Arcadia is associated with the wild goat god, with anything but civilization, urban living, or technology. It does not represent a solution to the conflicting demands of modern living, or any compromise between nature and culture or preservation and restoration' (1993, 21). This collision between multiple cultural philosophies and sf contexts does not so much recapitulate the pastoral image of Arcadia in its geographical specificity so much as it utilises and modifies formal pastoral associations within the framework of the sf text. De Luce's issue is more fundamental than a resistance to the use of the pastoral form, although she does see such use as politically reactionary. The problem of Turner's epic, in de Luce's view, is that its narrative of terraforming, of what she calls 'restoration ecology' after Turner's own term 'inventionist ecology,' provides no solutions or ways of thinking about contemporary ecological problems. She contends that '[w]hat takes place on Mars presents none of the challenges of restoration ecology in, let us say, the southeastern United States' (21). This perceived weakness perhaps accounts for Robinson's own insistence on the gains that terraforming Mars presents to Earth in his own *Mars* trilogy, but Turner's articulation of an image of the future, constructed from the tools of the past, need not spell out its application to real-world ecological issues.

Turner is concerned with exploring the shape of an epic future, not to provide solutions to contemporary issues, but to suggest a rich field of resources that can be utilised in constructing a heterogeneous global society. Older traditions allow time for the mistakes of the past to be revealed, revisited, and new routes mapped, routes that potentially avoid the historical problems associated with them. Writing a terraforming narrative is an artistic event that is concerned with a physical act of construction; it suggestively collapses the physical and the intellectual aspects of creation. As a way to tell the story of human evolution, the epic works as an interface between internal and external worlds. If epic tells the story of human evolution, sf tells the story of the possible ways in which we might evolve and develop, and it raises futures that can be sacrificed and disavowed, but which may be re-activated in new contexts. The poet's self-referential exploration of epic in *Genesis*, and the self-validating and self-referential practice of epic criticism that Turner engages in his critical work, explores fundamental issues in environmental philosophy, evolutionary anthropology, aesthetics, and other contemporary fields of enquiry that Turner shows to be relevant for the human project of re-territorialisation through landscaping.

5: Kim Stanley Robinson's *Mars* Trilogy

Kim Stanley Robinson's *Science in the Capitol* trilogy (2004; 2005; 2007) explores socio-political responses to climate change on Earth in a near-future setting, and features instances of geoengineering as a form of climate change mitigation. *2312* (2012) is set in a far-future solar system made habitable by a plethora of space habitats and terraformed planets, but it is with Robinson's groundbreaking *Mars* trilogy (1996c [1992]; 1996b [1993]; 1996a) that this study of terraforming ends. This trilogy engages in a dialogue over the cultural meaning of Mars, space exploration and terraforming. Its portrayal of Mars colonisation mirrors the structure of secessionist politics seen in such works as *Ecotopia*, which it uses to explore alternative socio-political arrangements. As Carol Franko explains, 'fiction is for [Robinson] the crucial realm for the human activity of asserting and testing values' (1997, 59). As in *Ecotopia*, the alternative practices experimented with in the new space of Mars begin to establish a feedback loop with the Earth of the text. One theme of the terraforming narrative is that of 'throwing together,' the combination of elements from many disciplines to establish the technical, social, political and economic basis for creating self-sustaining life-support systems on other planets.

The *Mars* trilogy incorporates overt environmental ethical reflection and brings this to bear against the industrial exploitation of Mars by multinationals that see it as a resource offering raw materials and a space for capitalist investment and development. Earth is forced to respond to the negative effects of climate change from a correspondingly global perspective, as the catastrophic effects of ecological disaster impact all the planet's population. The pastoral collectives that Bakhtin identifies are represented as nested collections of spaces, often associated with contesting positions that are developed throughout the narrative. The chronotope of the interplanetary mine in the *Mars* trilogy is connected to debates regarding the preservation of alien planets as wilderness areas, idylls where humanity's socio-political struggles have not yet reached. The growing politico-economic primacy of multinationals throughout the trilogy contrasts with the initial nationalist interest in terraforming: the colonists known as the First Hundred, although ethnically diverse, represent the joint – and paranoid – agendas of America and Russia. Later colonists from China, Japan, Switzerland and other countries, as well as various Arabic groups who derive their identity from religio-cultural sources (for example, Bedouin and Sufi), emigrate to Mars, introducing further nationalistic and cultural considerations to terraforming.

Jed Rasula has reconceived the intertextual aspect of American poetry as a metaphorical compost library. He argues that newness arises from the continual recycling of language, shaped by an author's attentiveness to predecessor texts and by reader interaction: '[i]n the compost library books have a way of collapsing into each other, not in the improvements of more "authoritative" editions or versions, but by constant recycling. Not one but many energies shape the field. It is a vortex' (2002, 17). Thierry Bardini argues that biological entities and processes are the ultimate junk, and that terraforming represents a prime example of composting for the creation of new forms and systems (2013). Bardini's sense of junk is rhizomatic; it is 'all kinds of stuff that grows in stacks and patiently waits for a renewed use' (2011, 7). Rasula and Bardini explore the questions of the ecological imperative of American poetry on the one hand and of junk as 'one of the signatures of this age' on the other (Bardini, 2011, 24), but both concepts, compost and junk, share this tendency towards exaptation in order to create newness in ways that add value.

Stephen J. Gould's term 'exaptation,' as Stuart A. Kauffman explains in his provocative *Investigations*, refers to the way in which the biological structures of Darwinian pre-adaptations evolve into new structures with capacities that, in many cases, could not have been pre-stated or predicted. Kauffman explains that 'in an appropriate environment a causal consequence of a part of an organism that had not been of selective significance might come to be of selective significance and hence be selected' (2000, 130). Metaphorical examples of this process of exaptation for literary purposes are compatible with a Bakhtinian view of language and with Broderick's view of the megatext, but they emphasise the specifically evolutionary and ecological character of this intertextuality. They can be linked to Brian Attebery's notion of the 'parabola,' a trajectory rooted in an iconic sf image that, appearing in a form subject to collaboration and jazz-like improvisation, is open to inventive variation: 'the sf scenario is an open curve, a swing toward the unknown' (2005, 14). The term joins this notion of a narrative trajectory to that of the parable, thus drawing attention to how the sf narrative 'combine[s] human interactions with scientific ideas and technological innovations in a meaningful way' (Attebery and Hollinger, 2013, viii).

Terraforming is a suitable motif for this view of the megatext as compost or junk: the emphasis that it places on the creation of soil, in stories such as Robert Heinlein's *Farmer in the Sky* (1967 [1950]) through to Michael Allaby and James Lovelock's *The Greening of Mars* (1984) and Robinson's *Mars* trilogy, establishes a connection to this notion of the fertility of the composting aspect of the library of texts – a fertility at once open and oriented towards the unknown. Percival Lowell's popularisation in the 1890s of his theory that the *canali* of Mars, identified by Giovanni Schiaparelli in 1877, were the traces of an irrigation system engineered by Martians to combat the resource scarcity of a dying planet adds another resonance to this notion of Mars and terraforming as a site for the composting of junk, that blend of romance and science that has informed the popular imagination of Mars since the late nineteenth century (Crossley, 2011, 73). Robert Crossley in *Imagining Mars* (2011) provides an excellent and extensive examination of the compost library of the Martian megatext and of the meaning of Mars as created by a complex relationship between science and the literary and popular imagination.

Robinson's *Mars* trilogy explores the fusion between the physical adaptation of the environment and the transformation of social practices and institutions. It considers the terraforming motif and its emphasis on closed life-support systems and soil, linking these physical parameters to an 'eco-economic' system propounded by the Martian colonists of the trilogy. Exploring how this system offers elements for exaptation from Earth's compost library of socio-economic and political practices and attitudes, this chapter considers the role of the Martian landscape as a distorted mirror of Earth that offers to transform and revitalise a planet consumed by tensions that exacerbate the global ecological crisis on a near-future Earth. Eric Otto discusses the trilogy's exploration of Aldo Leopold's 'The Land Ethic,' a classic work of environmental philosophy that proposes the extension of ethical consideration to non-human nature and which negotiates the space between science, economics, expediency and ethics (2003). Responding to Ernest J. Yanarella's criticism that the polyphony of subject positions in the trilogy allows Robinson to avoid resolving the ethical debate surrounding terraforming (2001), Otto argues that the work's multiple perspectives 'encourage readers to synthesize continually a complex array of political positions' (2003, 132). More recently, Otto has explored the ways in which environmental sf intersects with transformative environmentalism, that collection of environmental movements arising from the wake of Rachel Carson's 1962 *Silent Spring* (2012). These movements offer analyses of and solutions to environmental degradation, focusing, to use a cybernetic, ecological paradigm, on the inputs that feed into environmentally destructive behaviour, rather than on outputs that would require a reaction to specific examples of degradation (Otto, 2012, 1).

Carol Franko connects Bakhtinian dialogism, polyphony and the carnivalesque to elements of *Red Mars* (1997) and, while William J. Burling argues that Franko's insights cannot be usefully applied to the political process of *Blue Mars* (2005, 76), Robinson himself mentions in an interview that Franko offered 'a clear theoretical expression' of his aim that 'actually helped me in figuring out certain problems in *Blue Mars*' (see McVeigh, 1995, 4). In contrast, Burling argues persuasively for affinities between the political process outlined in *Blue Mars* and Ernesto Laclau and Chantal Mouffe's 'radical democracy,' points of contact that are coherent with notions of dialogism and polyphony and that extend Franko's analysis (2005). For Robinson, sf's environmental engagement possesses utopian dimensions: the *Mars* books are 'an attempt to take back the [utopian] territory – to show that the future is malleable and up to us' (see Buhle, 2002, 88).

Approaching Robinson's work from the perspective of composting is fruitful for exploring the connections between American ecological poetry and sf that he makes: as a former student of Gary Snyder and a dedicated burrower into a compost heap that includes American poets such as Thoreau and Emerson, along with Frederick Turner himself, Robinson states in an interview that 'I believe that science-fiction is one of the most powerful modes of poetry of all time. Science-fiction is just a metaphor for the world we live in and metaphor is one of the basic tools of poetry' (see Cooke, 1995). The Martian colony's attempt to establish a sustainable environment on Mars involves the creation of new myths exapted from experiences and systems of thought on Earth. In the context of the Martian environment, these narratives are tuned to the new specificities of a developing compost library. The creation of new myths on Mars enshrines a Martian narrative that offers the potential for feedback from the alien to the human in order to transform the multiple voices embedded in Earthbound history into new avenues for socio-cultural experimentation.

Gardens on Mars

Images of the garden appear as indicators of the drastic transformations that terraforming brings to Mars. In *Blue Mars*, Sax Russell is taken on a tour of Tyrrhena Massif near Sabishii, where the shaping of the Martian landscape prior to terraforming is described as '[n]ot chaos, technically speaking, but wild, speaking its unimaginable age in polyglot profusion' (Robinson, 1996a, 89). This polyglossia is influenced by the geologic and meteorological forces that have shaped the planet for billions of years. After his guides point out the colonising plant life, Russell understands that 'it was all fellfield, the whole Tyrrhena Massif,' an 'intensively cultivated' pastoral space (90). Strongly recalling Turner's depiction of Mars as a garden, Russell's guides explain that various cultural gardening aesthetics are deployed, ranging from the Japanese Zen of Muso Soseki and others to Fu Hsi's feng shui, the designs of Persian gardening gurus such as Omar Khayyam and the approaches of American ecologists such as Leopold, Wes Jackson and the biologist Oskar Schnelling. These voices serve as springboards for new visions of the landscape that have co-evolved with Mars's topology and climate, an ecopoietic technique that embeds humanity into the landscape through the co-adaptation of its gardening strategies with the land. Russell observes that the landscape is built from a polyphony of voices, 'an aesthetic journey, filled with allusions and subtle variants of tradition that were invisible to him' (91). These gardens represent the colonists' experimentation with the gardening practices of the past, a palimpsest of various aesthetic principles that speak of the legacy left to the Martians by Earth. These scenes are themselves part of the compost heap of the sf megatext, echoing Turner's depiction in *Genesis* of a garden on Mars built on the aesthetic practices of a multiplicity of cultures and stances towards the landscape. The soil for this riot of gardening techniques, however, is imported from Earth, illustrating another dimension to the colonists' continuing dependence on their home planet, despite the level of political and economic independence they have achieved (90).

For Russell's guides, ecopoiesis is 'terraforming redefined, subtilized, localized. Transmuted into something like Hiroko's areoformation' (Robinson, 1996a, 91). Moving from the global to the local, ecopoiesis re-visioned as gardening connects terrestrial landscapes to the primeval Martian planet, resulting in modifications to both. Like the areophany, this process is one of '[c]oevolution, a kind of epigenetic development' that subtly redefines the boundaries of the Martian land as an experiment in building homes that respect nature's otherness (91). While Russell initially views ecopoiesis as an unmediated process where the initial conditions are established by human agency ('let loose the seeds, then watch it all develop on its own. Self-organizing ecologies'; 92), Russell's guides see 'Mars [as] all a garden. Earth too for that matter. This is what humans have become. So we have to think about gardening, about that level of responsibility to the land. A human-Mars interface that does justice to both' (91). Echoing Sargent's *Venus* trilogy, this view of terraforming as gardening is an interface that mediates between human communities and non-human nature.

At Shining Mesa, along the banks of a stream fed by meltwater from the Marineris canyon floors, 'forest galleries were springing up' whose 'balsa canopies were allowing a great number of plant and animal species to flourish underneath them' (Robinson, 1996a, 379, 380). According to Nirgal's acquaintances, 'it was the most diverse biotic community on Mars' (380). This landscape signals the extent of the transformations Mars has undergone after colonisation. Nirgal's tour of Shining Mesa takes place alongside an internal struggle over two conflicting modes of inhabiting the land – between the nomadic lifestyle he is familiar with and the settled life of a home: '[h]e wanted to live in the open air. To learn a patch of land, its soil and plants and animals and weather and skies, and everything else' (382–83). Candor Chasma is too sublime a location, too much a wilderness, for individuals to establish roots and inhabit the land. Nirgal therefore decides to resume his search for the absent Hiroko, thus allowing the narrative to explore the physical changes to the Martian landscape on the now blue Mars.

Nirgal's view of the Elysium Massif strait is like nothing he has ever seen before: 'water, the sea, a whole future world' (Robinson, 1996a, 390). During this search he realises that the familiar primeval Mars is gone. His childhood companions encourage him to stop looking for Hiroko, and he decides instead to 'look at the land,' prompting a return to Tyrrhena Massif to try his hand at gardening (396). Thus, he moves from a nomadic to a settled lifestyle: '[h]e would be an ecopoet,' but '[f]irst he had to learn the land' (398, 399). Nirgal becomes the steward of a small basin, where 'working out one's locality's connections to the larger region was a big part of the ongoing process of ecopoesis [*sic*]' (400). As he cultivates his allocated basin, a microcosm of the plant and animal life that continues to be established at Tyrrhena Massif, a 'Tyrrhena committee on the introduction of new species' visits Nirgal in order to establish a position towards the local ecology: 'there was a growing sentiment to regard this mix as "natural" to Tyrrhena, to be altered only by consensus' (405). After this visit, Nirgal says to the marmots who have taken up residence by his home, 'now we're indigenous': on Mars, being indigenous is a matter of political consensus (405).

Another perspective on the transformations to Mars is supplied by Ann Clayborne, whom Russell accompanies on an expedition and observes reading the landscape like a text. Russell sees her as an oracle, a visionary whose love of rock – of abiotic life – he compares to Hiroko's visionary areophany. He attempts to uncover the etymology of the words 'stone' and 'rock', but quickly abandons himself to a Mars landscaped as a tabula rasa, a seemingly boundless space invested with a creative potentiality directed towards the future. Such a view contrasts with his thinking about the etymology of the word 'garden'. Derived from the Old Norse *gard* for enclosure, it '[s]eemed to share origins with guard, or keeping' (Robinson, 1996a, 92), thus resonating with the image of the dome as a protective enclosure. The rocky Martian land is a space opposed to the cultivation of gardens seen in Tyrrhena Massif and the banks of the Arena Glacier: '[w]ithout active gardening, this was what one got' (100). The encroaching plant life is a source of anger and depression for Clayborne, who interprets it as the destruction of Mars. Michel, a psychologist and member of the First Hundred, encourages her to walk amongst the changed Martian landscape and argues that terraforming has made humankind a part of the land and its ecology, essentially embedding them into a new home. For Clayborne, Michel suggests, '[*y*]*our task becomes seeing the Mars that always endures*' (257).

Encouraged by this suggestion, Clayborne embarks on a walkabout to rediscover her relationship to the now altered Mars. Her encounter with the new landscape brings her face to face with an ecology imported from Earth, the short food chains of the Antarctic comprising marine, avian and mammalian life. Along with the image of skuas scavenging a dead seal, Clayborne sees a polar bear, which gives chase to her after feeding from the same carcass. This pivotal encounter introduces a new vantage to her consciousness: '[s]he only had to close her eyes and she saw again that heraldic image of the bear flowing over the rock; but open them and there the dashboard gleamed, bright and artificial and familiar. Ah so strange!' (Robinson, 1996a, 268). For Clayborne, the image of the polar bear is an emblem for change: it is a haunting vision that initiates a transformation that eventually allows her to step back from her nihilism, preparing her for a modified relationship to the new Mars.

'Stepping Back'

The environmental philosopher Holmes Rolston III identifies two broad categories of worldview around which debate over the aesthetic appreciation of nature revolves: 'one that aesthetic experience must be participatory, relating an actual beholder to a landscape; the other that nature is objective to such beholders, actually known in the physical and biological sciences' (1995, 377). Considering the question of whether aesthetic appreciation of landscapes needs to be science-based, Rolston compares those various and diverse examples of what he calls 'prescientific' approaches to nature which, on the one hand, characteristically misunderstand nature (from the point of view of scientific objectivity), while on the other have developed specific ways of relating to that nature based on human co-adaptation with the environment. Rolston reflects on the assumption that 'no one appreciates the canyon, for what it really is, unless helped by geologists' (374) and argues that '[s] cience cultivates the habit of looking closely, as well as of looking for long periods of time. One is more likely to experience the landscape at multiple scales of both time and space' (376). Nevertheless, he is well aware of the problems of contingency associated with this claim, noting that 'science or no science, everyone can gain some of that sensitivity' (377).

Kauffman sees the relationship between science and story as one in which science itself can be made richer by incorporating storytelling into its practice. He argues that '[t]he propagating exapting biosphere is getting on with it, and it appears that we crucially need stories to do some of the telling of that getting on with it' (Kauffman, 2000, 135). Kauffman explains that '[s]tories are our mode of making sense of the context-dependent actions of us as autonomous agents. And metaphor? If we cannot deduce it all, if the biosphere's ramblings are richer than the algorithmic, then metaphor must be part of our cognitive capacity to guide action in the absence of deduction' (Kauffman, 2000, 135). Thomas Heyd suggests that the aesthetic appreciation of nature need not rely on science and that science may actually be harmful to such appreciation because it directs attention to the 'theoretical level and the general case,' rather than to 'the personal level and the particular case that we actually need to engage' (2001, 126). Heyd argues that the 'aesthetic appreciation of nature should be guided by a great variety of stories from a diversity of walks of life and cultures because this enriches our capacity to appreciate nature aesthetically' (137). Interplanetary imperialism, a tradition substrating such classics as Ray Bradbury's *The Martian Chronicles* (1958 [1950]) and Heinlein's *The Moon Is a Harsh Mistress* (2001 [1966]), has traditionally interpreted cosmological nature as a field for American conquest and annexation. Sf has simultaneously challenged this imperialism, a trend that feeds into Robinson's exploration of human–nature relations in his terraforming trilogy.

Robinson's *Mars* trilogy portrays a not unfamiliar Earth subject to overpopulation, pollution, global warming, rising sea levels, war, famine and severe economic and political inequalities between the rich and poor, a list that matches Glen Love's list of contemporary environmental problems (2003, 14–15). These ecological changes effect social, political and economic repercussions. In *Red Mars*, a group of one hundred scientists known as the First Hundred begin terraforming Mars in 2027. The text foregrounds the importance of, and problems associated with, opening spaces for dialogue in two ways: first, by utilising different perspectives to focalise each section and, second, by portraying the social activity of the colonists from these alternating perspectives. This involves depiction of the confrontations and differences of ideology between opposed groups or individuals. The Machiavellian Frank Chalmers recognises the importance of exchange between opposed positions when orchestrating his rival's assassination in *Red Mars*'s opening sequence. He thinks of his chosen assassin as a 'fool [...] talk means everything. We are nothing but information exchange, talk is all we have!' (Robinson, 1996c, 31). Chalmers's plan for political dominance involves sowing discord between cultural groups. By manipulating the historical antagonisms between Arab and American nationalism, Chalmers sabotages any attempt towards dialogue. Raising awareness of and support for political positions and lifestyles is a major narrative component and is conducted through discussion, campaigning and interviews, further reflecting the importance of information exchange. As multiple individuals and groups attempt to define the Martian landscape according to their own values, the exchange and conflict between their contesting positions establish a literary ecology that opens a space for debate over the meaning of Mars.

Central to the *Mars* trilogy is the depiction of a developing Martian community and its struggle for political independence from Earth. One view of Mars is purely instrumental in the sense that it is seen only as a means to relieve the pressures threatening Earth's population. In response to uprisings on Mars the official line from Earth's governments is that '"Mars is not a nation but a world resource"' (Robinson, 1996c, 602). This perspective is directly connected to the mining that had begun earlier in Antarctica, a space that, as a natural reserve, has been protected from such activities by the 1961 Antarctic Treaty. Clayborne places the blame for the breach of this treaty squarely on the colonists' efforts to terraform Mars: she explains that '[t]hey kept mining and oil out of Antarctica for almost a hundred years [...] [b]ut when terraforming began here it all collapsed' (298).

This view of nature as a resource highlights technology's influence in redefining humanity's treatment of external space. The chronotope of the first colonising outpost in *Red Mars* implies a series of narrative trajectories that draw upon the pastoral opposition between images of the 'natural' landscape of the country and the technological city. As Bakhtin explains, the chronotope is the artistic representation of space and time as interrelated in a text; time qualifies spatial meanings and vice versa. This representation, because linguistic, allows it to accrue a series of human-centric meanings from structures internal to the text and through sf's megatext, a repository of discourse constructed by works in the sf tradition and by reader interaction with those works. Russell's assessment of terraforming is cast in doubt in the light of Nadia Cherneshevsky's trip with Clayborne to the north polar region. On her return, she sees their habitat in a new light: '[i]t had the disordered, functional, ugly look of Vanino or Usman or any of the Stalinist heavy industry cities in the Urals, or the oil camps of Yakut. They rolled through a good five kilometers of this devastation' (Robinson, 1996c, 191). The description of the outpost taps into the narrative potential of the dystopia, the alternative to the possibility of a utopian interplanetary colony. This theme contributes to the ongoing debate regarding the development of new societies on other worlds. It constructs an image of a repressive society signified by the chronotope of the city as wasteland, delineating a socio-political structure that rejects heteroglossic dialogue that would incorporate all the voices of the multiple groups who work towards constructing a new Martian identity.

Greater control of the environment is thus accompanied by a greater willingness to adapt the landscape for purely anthropocentric ends: nature's otherness does not feature as a constraint to terraforming. Cherneshevsky responds to the environment instrumentally, although this response is personal rather than economic. When Clayborne invites her to a trip to Mars's polar regions, she sees the wilderness of the Martian landscape and experiences a cognitive shift that allows her to re-evaluate this space. She thinks that '[a]ll this beauty was so strange, so *alien*,' and the narrator continues: 'Nadia had never seen it properly before, or never really felt it, she realized that now; she had been enjoying her life as if it were a Siberia made right, living in a huge analogy, understanding everything in terms of her past' (Robinson, 1996c, 171). Recognising the alien aspect of Mars's landscape helps her identify her own tendency to project meaning onto nature and so ameliorate its otherness. Cherneshevsky's identification of Mars with Siberia instantiates a landscape that is completely constituted by her imposition of meaning. Her initial view of Mars as a new Siberia is instrumental because it allows her to cope with the deterritorialisation that accompanies the unfamiliarity of, and the demands of living within, an alien landscape.

Here we see the pastoral refigured: by contrasting the alien with the familiar, its strangeness is raised to the level of awareness, drawing attention to how Mars is unlike familiar natural landscapes. It signifies the new; civilisation on Earth is shifted to a nostalgic past. Cherneshevsky responds to the intuition that Mars is not solely a field for the imposition of her engineering discipline, thus overlaying the chronotope of the Martian wilderness with non-instrumental value. This episode draws on the mystique of Mars, constructed, as Russell notes, from '[a]ll those dumb sci-fi novels with their monsters and maidens and dying civilizations. And all the scientists who studied the data, or got us here,' but it challenges conventional ideas of beauty, pushing us as readers to reflect on a personal, aesthetic response to a nature not already conceived of as determined by instrumental cultural expectations (Robinson, 1996c, 212). Through Cherneshevsky's experiences of the 'real' Martian landscape, an ecologically oriented perspective is woven into the dialogue of the text and stands out as one of its major ideological voices.

Cherneshevsky's approach is indicative of general trends towards landscaping Mars. Hailwood focuses particular attention on what he calls 'a kind of intellectual fragility involved with the difficulty of maintaining natural otherness in view: the ease with which it is overlooked in the cultural process of feeling at home and secure in a landscape' (2004, 29). Earth's view of Mars as a resource exemplifies a landscape that prefigures and justifies physical terraforming. Clayborne reflects that '[o]nly on Mars did they walk about in a horrendous mishmash of the dreams of the past, causing who knew what disastrous misapprehensions of the real terrain' (Robinson, 1996b, 155). Such 'disastrous misapprehensions' disrespect nature's otherness and demonstrate the failure to recognise the constructivism inherent in landscaping. Clayborne and Cherneshevsky are both aware that such cultural projections are ways in which continuity with the past is maintained. An awareness of intellectual landscaping processes allows Cherneshevsky to acknowledge that physical space is not solely constituted by anthropomorphic perspectives.

Franko argues that intersubjectivity is central to two of Robinson's short stories (1994), 'Exploring Fossil Canyon' and 'Green Mars,' collected in the *Mars* trilogy's companion volume *The Martians* (2000 [1999]). They 'explore the subjective processes that shape such debates [about nature's value] and suggest that the most important moments of growth are those that involve a crisis in one's perception of otherness, and that such crises open the way for the discovery of a social utopian impulse, one that responds to otherness, human and non-human' (Franko, 1994, 199).

'Instrumental' is a term from environmental philosophy that refers to the way in which we value nature, and it forms one of the dimensions of the concept of nature's otherness. Hailwood argues that 'the value conferred by nature's otherness is best thought of as non-instrumental (independent nature is a "negative end" [a constraint] in virtue of its otherness), extrinsic and objective' (2004, 13). Its value is located 'extrinsically' by virtue of the fact that it is other to humanity. Nature's otherness also has objective value, and here Hailwood adopts Thomas Nagel's conception of objectivity, which he describes as 'a method of understanding from a detached perspective, formed by stepping back from an initial view to arrive at a new conception taking in the original and its relation to the world' (52).

In this light, Bakhtin's notion of transgredience, the practice of assuming an external perspective as a mode of consciousness, can be considered a form of objectivity in the sense that Nagel describes. Evaluating the self from the perspective of nature's otherness offers an avenue for overcoming anthropocentrism and for developing new ethical relations with nature (Murphy, 2011, 156). Cherneshevsky's cognitive shift is an example of this process of 'stepping back' and reconsidering one's relationship to the environment. There are other implications involved with this conception of objectivity: Hailwood cites Nagel's explanation that '"[t]he wider the range of subjective types to which a form of understanding is accessible – the less it depends on specific subjective capacities – the more objective it is,"' but notes that, to avoid nihilism, '[n]ormative realism [...] requires the retention of some relatively subjective element' (2004, 52). Objectivity is therefore dependent upon the convergence of multiple subjectivities, yet this process exists in tension with individual subjective perspectives that allow individuals to evaluate and develop a range of new ethical relationships to nature and to regulate a potential regress towards nihilism.

The debate between the Reds and Greens is fundamental to the trilogy and is represented by the opposition between the geologist Clayborne and Russell, the physicist turned biotechnologist. In *Red Mars* they confront each other and establish their initial positions towards terraforming. Clayborne argues that 'you're going to wreck the historical record, destroy the polar caps, and the outflow channels, and the canyon bottoms – destroy a beautiful pure landscape, and for *nothing at all'* (Robinson, 1996c, 212). While Clayborne's emphasis on the physical fragility of natural features apparently supports an intrinsic view of nature's otherness, her discussion of the geological 'historical record' and an aesthetic of beauty suggest that this is otherwise. Russell, on the other hand, argues that

> Changing it won't destroy it. Reading its past might get harder, but the beauty of it won't go away. If there are lakes, or forests, or glaciers, how does that diminish Mars's beauty? I don't think it does. I think it only enhances it. It adds life, the most beautiful system of all [...] Mars will always remain Mars, different from Earth, colder and wilder. (Robinson, 1996c, 213)

Russell places value on aesthetic aspects of the landscape, but for him only biotic processes – organisms and ecologies – have extrinsic aesthetic value. He recognises the presence of otherness as a component of nature on Mars and he distinguishes nature's otherness from an alien nature's otherness. Clayborne is dissatisfied with Russell's more popular position but is unable to effectively articulate a response, demonstrating the difficulty involved in speaking for the rights of abiotic nature. She does, however, attempt such a defence:

> I think you value consciousness too high, and rock too little. We are not lords of the universe. We're one small part of it. We may be its consciousness, but being the consciousness of the universe does not mean turning it all into a mirror image of us. It means rather fitting into it as it is, and worshipping it with our attention. [...] You've never even seen Mars. (Robinson, 1996c, 213–14)

Clayborne advocates an attention to nature's otherness that is almost religious, criticising the anthropocentric landscaping that follows from the view of humanity as 'lords of the universe.' She sees terraforming as imperial in its approach to nature when she completes Russell's claim that '[d]eciding to go to Mars is like the first phrase of a sentence, and the whole sentence says—' by saying '*Veni, vidi, vinci*' (Robinson, 1996c, 56). This 'sentence,' the narrative of interplanetary colonisation that seems to demand resolution in familiar ways, is the political landscape that informs terraforming. This position is distinct from her earlier emphasis on aesthetic beauty and is intrinsic in the sense that the value of nature's otherness is not dependent on its relationship to humankind. It is nevertheless extrinsic in that an appropriate response to the universe is one of paying attention and of 'fitting into it as it is,' which does depend on recognising it as other from humanity's perspective. Clayborne's view in this pivotal episode involves extrinsically valuing landscapes in the manner of Hailwood's notion of nature's otherness. Robinson's *Mars* trilogy attempts to negotiate new relationships to a nature already intellectually landscaped with meaning.

This relation of extrinsic worth is confused, however, because Russell refers to Clayborne's position as advocating intrinsic worth, as when he thinks that 'she believed in some kind of intrinsic worth for the mineral reality of Mars' (Robinson, 1996b, 186). Russell fails to understand Clayborne's arguments; *Green Mars* and *Blue Mars* are partly concerned with his decision to step back from and re-evaluate his initial position, partly to better understand Clayborne's own perspective and partly to attempt to convince her that the terraforming project has its own kind of value. On a walking trip with Clayborne in *Blue Mars* he realises that '[o]ne had to let things speak for themselves. This was perhaps true of all phenomena. Nothing could be spoken for. One could only walk over the land, and let it speak for itself' (Robinson, 1996a, 98). If, however, it is not possible to speak for the landscape, how do you convince those who disrespect nature's otherness to maintain a green (or rather red) perspective if not by relating its value in some way to humanity? Accepting nature's otherness as an extrinsic value is one way in which characters begin to recognise existences external to humanity. Clayborne's advocacy and, eventually, Russell's focus on viewing the landscape open up other textual spaces where appropriate responses to the environment are explored.

The Red/Green positions splinter into a variety of mediations between these two extremes, and Russell comes to question his initial orientation. Setting the narrative on the red planet allows the ideological connotations of 'green' as a label for environmental consciousness to be inverted, with it coming to represent the interests of an unreflecting and destructive process. That these values have led Earth to its current environmental and political crisis allows us to question further our notions of an acceptable interplanetary environmental consciousness. These values also lead us to consider the symbolic value of terraforming and to explore the implication that terraforming leads to a mirroring of the socio-political dynamics on Earth. Space is thus politicised as groups struggle to define the meaning of Mars and their relationship to it.

Clayborne and Russell exemplify the way in which language is used to speak for the Martian landscape from perspectives that view it as a site of traditional symbolic value and from contesting positions, contributing to the definition of textual spaces for the confrontation and interaction of different discourses. Russell views Mars as lifeless and therefore ripe for the seeding of life, a traditional sf theme. Clayborne sees terraforming in terms resonating with the American pastoral, with the Martian landscape occupying the role of pristine wilderness. These values are represented synchronically, as spaces that are placed in juxtaposition to each other: Clayborne and Russell embody contesting views towards the Martian landscape, while Chalmers considers it a space in which to manoeuvre for political dominance. However, through the diachronic structure of a text (narration and character dialogue), these and other positions come into contact with one another to allow an implied audience to consider and question the value systems represented. Different characters may express contesting positions towards a particular landscape that can be read against the implied authorial voice. As Russell's reassessment of his initial position shows, characters may be exposed to different ideological worlds in their attempt to reconcile the discourses that these landscapes are made to represent. Confrontations with nature's otherness deterritorialise and thus allow characters to step back, enabling them to develop alternative ways of valuing their environment.

Visions Reflected Back to Earth

While visiting the Alps Nirgal experiences the realisation that 'Earth was so vast that in its variety it had regions that even out-Marsed Mars itself – that among all the ways that it was greater, *it was greater even at being Martian*' (Robinson, 1996a, 201). This impression is generated by the similarities and differences of climate between the Alps and a terraformed Mars, one in which Nirgal, as one of the first children born on Mars, feels at home in. Antarctica, as we learn in the opening of *Red Mars*, is also *'a landscape that was almost as cold and harsh as Mars itself,'* which explains why it functions as the selection and training site for the First Hundred prior to their journey to Mars (Robinson, 1996c, 41). What this relation highlights is the presence of nature's otherness in all landscapes. Even on Earth, where many spaces are significantly modified by physical and psychological cultural adaptations, an element of this otherness remains, reminding the reader that human ends and values do not exist in isolation from the natural world. Nirgal is Martian and so Earth for him represents an alien planet; this perspective offers the reader an opportunity to engage in Nagel's process of stepping back and reassessing their subjective view of nature's otherness in the familiar landscapes of Earth.

The *Mars* trilogy argues that a reliance on technological fixes as an answer to societal conflict is inadequate. Championed by the Russian Arkady Bogdanov and the American John Boone, a break with history becomes one of the driving goals of the more politically minded on Mars. They believe that the social patterns on Earth are responsible for many of the ills that humanity faces and that nothing but a complete overhaul will do: a discarding of those that are unhelpful and destructive and a retaining of those that speak for cultural pluralism and a new Martian identity that offers the colonists a global identity distinct from Earth. Boone, after many discussions with Bogdanov, advocates 'a new Martian way, a new Martian philosophy, economics, religion!' (Robinson, 1996c, 410). This call for a new outlook alludes to Isaac Asimov's short story 'The Martian Way' (1974 [1952]), and is one of the central ideas of sf narratives of terraforming. This theme takes its place in an ongoing debate that addresses issues of economic and socio-political independence on Earth. This economic independence is linked to a Martian perspective where local forms of community, along with the appropriate forms of cultural and economic interchange between communities, are emphasised in contrast to Earth's socio-political schisms, themselves caused by individualistic and nationalistic perspectives towards resources and land use. Terraforming literalises metaphors for the creation of discursive spaces to explore new forms of local and global connectedness and identity that stand as alternatives to destructive social formations on Earth. The intersections between texts and between generic categories such as sf, utopia and the pastoral are heteroglossic and engage with the sf megatext and with contemporary discourses of environmental philosophy and geopolitics.

That the initial colonising outpost is built near resources deposited from Earth, including a full range of technologies built by Boeing, Rolls Royce and other companies, indicates that the outpost itself and the terraforming effort that it supports are driven by the commercial interests of the most powerful of Earth's transnationals, a political fact that some of the First Hundred, unlike Bogdanov, would prefer to ignore: 'it all comes back, and we have a return of ownership, and prices, and wages. The little scientific station is being turned into a mine, with the usual mining attitude toward the land over the treasure' (Robinson, 1996c, 403). This draws on the sf megatextual trope of technological sophistication, but associates these technologies with familiar companies, implying one of the uncanny oscillations of the subjunctivity of sf, which Samuel R. Delany claims is 'blanketly defined by: *have not happened'* – an unspoken 'yet' may linger at the end of that sentence (1977, 44).

The chronotope of the scientific station, associated with exploration and the scientific utopia, is shifted to that of the mine, an industrial, capitalist image. Driving these economic interests is the application of advanced technology as a means of securing the resources to relieve scarcity on Earth, thus allowing Earth's governments to cope with the growing ecological crisis. The historical application of increasingly advanced technology has enabled Earth's population to boom, resulting in a depletion of natural resources and an amplification of the growing crisis. The Martian discovery of a treatment that can significantly prolong human lifespans further emphasises this dynamic. Some predict that this will increase the transnationals' drive for economic security for two reasons: because it will exacerbate the already problematic division between the rich and poor, and because it will increase overpopulation on Earth: 'if this damned treatment only goes to the rich, then the poor will revolt and it'll all explode – but if the treatment goes to everyone, then populations will soar and it'll all explode' (Robinson, 1996c, 415). Mars is seen only as a space for the resolution of Earth's ecopolitical problems. The view of the scientific station and of Mars as a mine represents the transnationals' attempt to impose an identity based solely on utilitarianism onto the Martian landscape and its community. Because these transnationals see Mars only as a source for the extraction of resources, and because they view this as the scientist's only role, their priorities for the physical adaptation of the landscape reflect an attempt to turn the planet into a mine. The scientific community on Mars is considered only in utilitarian terms, not as a community whose lived experience on the planet might legitimate alternative interpretations of the landscape.

The urge to make a break from the trajectory of reified ideology is mirrored by the narrative's discontinuity in time. Beginning in 2026, this near-future narrative compresses the traditional gap between the time *Red Mars* was published (1992) and the far-future setting of much sf dealing with the colonisation of the solar system. Such far-future narratives imply significant changes to socio-political structures and technology, as Sargent's *Venus* trilogy illustrates. The *Mars* trilogy begins by retaining many of the structures that are now familiar and calls them into question as the narrative progresses. It also adds further weight to the notion that the series of crises faced on Earth oscillates between Delany's categories of sf subjunctivity, 'have not happened' and 'have not happened yet,' but may soon (1977, 44). The solution demonstrated by the text's emphasis on social relationships in a new experimental space asserts that it is not by focusing on adapting the landscape but by landscaping the self, by metaphorically terraforming the individual and social aspects of a community, that the best hope for effective change is to be realised. This is wrapped up with the Martian landscape: a phrase that appears throughout the trilogy, '[s]o we terraform the planet; but the planet areoforms us,' demonstrates that Mars has a corresponding effect on the identity of its inhabitants and suggests that there is an influence exerted on this sense of planet from an alien nature's radical otherness (Robinson, 1996c, 301).

Sf taps into a range of generic forms and discourses and reconfigures them through its own language in order to provide textual spaces for examining human relationships to the landscape. As shown in the foregoing discussion, this is achieved by adapting the pastoral and utopian form in conjunction with portraying debate between positions to consider questions of land use and responsibility towards the land. These alternative perspectives arise from differences in physical space, from the contrast between the planetary spaces of Earth and Mars and the meaning invested in them. Local, global and interplanetary space is constituted by ecologies of landscapes embodying multiple ideological positions. In this way the concept of the chronotope, as a unit for the analysis of texts, is joined to Bakhtin's theory of dialogism, which is especially important as it is the principle by which the megatext operates. Dialogism and the chronotope interact to define the structure of a text; parts interrelate to take on additional layers of meaning, and changes to one dimension impact upon others. In the glossary of *The Dialogic Imagination* dialogism is described as 'the characteristic epistemological mode of a world dominated by heteroglossia': '[e]verything means, is understood, as a part of a greater whole – there is a constant interaction between meanings, all of which have the potential of conditioning others. Which will affect the other, how it will do so and in what degree is what is actually settled at the moment of utterance' (Bakhtin, 2002, 426). The Red/Green debate and the view of technology as the solution to ecological problems on Earth are specific examples of voices that work to condition each other. Linked to the internal ecology of a text is an external one where individual works exist in dialogue with others, thus contributing additional layers of meaning to the text's themes and images.

Closed Life-Support Systems, Soil, and Cybernetics

Biospheres are materially closed systems that are energetically open to the influx from the sun. They are microcosms of larger ecological systems on Earth and are intended to function as models for the Earth as a global system. Martyn J. Fogg draws comparisons between contained biospheres such as Arizona's Biosphere 2 and the uncontained biosphere of Earth, arguing that the ultimate aim of terraforming would be to create a self-sustaining uncontained biosphere on another planet (1995). These uncontained biospheres rely on the physical cycling of elements within a closed system, such as hydrological cycles, various biotic cycles and nutrient cycling through several processes. In *Red Mars*, the scientific team responsible for agricultural work tinker 'endless[ly with the] project of maximizing the closure of their biological life support system,' the success of which is measured against a formula, $K = I - e/E$. The formula itself, closure equals Hiroko's constant minus the rate of incomplete closure divided by the rate of consumption in the system, expresses the simplicity of the closed cycle, the ideal of which would be $K = I - 1$, or closure equals Hiroko's constant minus 1. Since Hiroko's constant is a fictional term, this equation is important not for its scientific veracity but as an ideal with powerful metaphorical implications for the Martian colonists. The ideal goal is acknowledged to be 'unreachable, but asymptotically approaching it was the farm biologists' favorite game, and more than that, critical to their eventual existence on Mars' (Robinson, 1996c, 85).

Physical life-support systems connect with others in relationships whose complexity is compounded by cultural elements that occupy multiple dimensions within a system of subsystems, all of which are open. Since constituents of life-support systems include cultural aspects, and as no system can obtain complete closure – Earth is energetically open to sunlight and radiation from the solar system and leaks gases into space – physical and cultural systems retain a capacity for externally influenced transformation. Incomplete closure in this context is tied to utopian thought, which Robinson redefines as a process involving continual change and not a static blueprint. Robinson reflects that 'Joanna Russ talks about changing the term from Utopia to Optopia, meaning "the optimum possible" – a continuous, dynamic process. Even HG Wells in his Utopian writing would often talk about this kinetic process rather than reaching any kind of stasis' (see Cooke, 1995). This formulation chimes with one aspect of Rasula's discussion of the compost library, which resists closure and continually re-activates Palaeolithic lore and historical voices in new contexts (2002). The propensity for the sf parabola to offer a binocular vision that 'allows us to view stories from two perspectives at once, as both literal description and metaphor,' likewise resists closure through its vacillation between modes of meaning (Attebery and Hollinger, 2013, ix). Before further developing this connection between the terraforming motif and the compost library, it is necessary to consider the implications of the motif of compost and soil as it figures in Robinson's terraforming trilogy.

The biotic modification of Mars, otherwise known as ecopoiesis, borrows from the insights of Lovelock's Gaia hypothesis, specifically the notion that organisms and their environment are involved in feedback systems that regulate the environmental parameters of a planet. This feedback system, understood as the institution of a life-support system, must be managed on Mars. '[*N*]*ew life fed on the compost of their ancestors, and reproduced again. Lived and died; and the soil and air left behind were different than they were before these millions of brief generations'* (Robinson, 1996c, 245). Life is involved in a bootstrapping process in which organisms rely on the compost of their ancestors in order to thrive and so change their environment. This image possesses a metaphorical parallelism when 'life' is extended to include the colonists themselves, whose own ability to modify Mars builds upon the 'compost' of their own ancestors. Bardini defines computer pioneer Douglas Englebart's use of the term 'bootstrapping' as 'an iterative and coadaptive learning experience,' a notion that grows out of Norbert Wiener's influential theorisation of cybernetics (2000, 24). In *Junkware,* Bardini undertakes a (bio)semiotic examination of junk DNA in terms of 'an inquiry into the cybernetic metaphor applied to the understanding of life, its modes of reference, and the question of "genetic insignificance"' (2011, 21). 'Junkware,' Bardini explains, 'is the name I chose to give this ordeal, turning the modern industrial and postindustrial excretions into a new sense of what being human can mean, now' (24). This aim resonates with that of the sf parabola, which builds on icons that are exapted in ways that vacillate between literal and metaphorical signification without offering definitive resolution.

Lovelock's view of Earth as a Gaian system builds on cybernetics and exploits the potential for analogies to be drawn between domains implicit in systems theory and its probabilistic approach to processes shared by a variety of structures. Rasula explores the metaphor of compost as a figure for a geographically bounded intertextuality exemplified by American poetry of the Black Mountain school, which he notes 'was historically congruent with, and sometimes affiliated with, the interdisciplinary matrix gathered around what Norbert Wiener named "cybernetics"' (2002, 3–4). Such intertextuality hinges on what he calls the biodegradable, transformative potential of language, and specifically of the trope:

> In the tropics of American poetry, trope is the composting engine, a fundamental dislocation, forge or furnace of a different locus: the unpropertied space germane to language. Not the mysticism of another world, but another economy (another *oikos* or household) of language-in-production, words in emanation, not nation. A tropical poetry is an agency of partial bodies, effluvia, surplus meaning: partial to polysemy, many-seeding. (Rasula, 2002, 124)

A focus on language is fundamental to Wiener's conception of cybernetics, concerned as it is with both communication and control acting on a system's capability to generate feedback and so achieve homoeostasis. Wiener explains that '[i]n control and communication we are always fighting nature's tendency to degrade the organized and to destroy the meaningful; the tendency, as Gibbs has shown us, for entropy to increase' (1988, 17). Rasula's description of another 'economy [...] of language-in-production,' of a 'tropical poetry [...] partial to polysemy' and imbued with a transformative potential associated with its locus as an unfamiliar or alien space, can be brought to bear upon Robinson's treatment of the habitation of an alien planet and the development of living practices that are tied to place (2002, 124). The *Mars* trilogy narrates a colony's struggle to bootstrap and develop complexity in variety and structure as it terraforms and learns to inhabit the planet. The trajectory of this bootstrapping process is structured like an sf parabola, as a movement towards the unknown and as an ecological parable. The sf trope, the motif of terraforming, is itself a composting engine in which a variety of domains of knowledge collide and are transformed.

Scientific discourse is applied to narrative considerations of soil, which sustains important nutrient cycles that determine the potential and the character of the organisms that are able to take root on Mars. '[D] ifferent soils encouraged or discouraged each cycle to different degrees' (Robinson, 1996a, 340); understanding the ways in which micronutrients like 'iron, manganese, zinc, copper, molybdenum, boron, and chlorine' work, along with macronutrients such as 'carbon, oxygen, hydrogen, nitrogen, phosphorus, sulfur, potassium, calcium, and magnesium,' becomes essential for the terraforming effort (339). Given the complexity of soil composition, the polyphonic range of soil types and the geologic span of time that it takes for organisms to create soil, the terraformers realise '[t]hey were going to have to construct soil just like they had the magnesium bars' (Robinson, 1996c, 140). Despite the sophisticated technologies that they have at their disposal – space elevators, solettas (giant mirrors in orbit that reflect sunlight to the surface) and the like – 'manufacturing soil was one of the most difficult technical feats they had ever undertaken' (Robinson, 1996a, 338). Much of the narrative shifts towards issues of compost, agriculture and gardening, thus allowing a focus on landscape to dominate the text. The compost/soil motif is one of the core images of the *Mars* trilogy; it is redefined as polyphonic in terms of its vast range of types, which in turn support a range of different organisms and, with regard to the colonists, a multiplicity of lifestyles. It is both a literal life-support system for the colonists and a model that functions as a parable for productive cultural and political variation and diversity.

Eco-Economics and the Landscape as Mirror

Speaking of his alliance with a formulation of scientific socialism that rejects nineteenth-century scientism, Robinson claims that

> the 'scientific' returns as a way of talking about providing some kind of ecological basis to economy. That way, economy is not just the astrology of the ruling class but actually a way of calculating true costs and benefits in a way that could be agreed on and quantified, and therefore making clear what we are really doing – whether it is sustainable or not over the long haul. (See Buhle, 2002, 89)

It is this dislocation and movement towards another economy, one of creation rather than depletion, that underlies Boone's call for seeing the terraformation and habitation of Mars in terms of an eco-economic system in a speech he delivers to his fellow colonists in *Red Mars*:

> 'Look,' he said, 'here we are on Mars!' (Laughter) 'That's our gift and a great gift it is, the reason we have to keep giving all our lives to keep the cycle going, it's like in eco-economics where what you take from the system has to be balanced by what you give in to it, balanced or exceeded to create that anti-entropic surge which characterizes all creative life and especially this step across to a new world, this place that is neither nature nor culture, transformation of a planet into a world and then a home.['] (Robinson, 1996c, 443)

A revolution for independence from Earth becomes the main narrative trajectory of the trilogy, and eco-economics becomes the basis from which this revolution is conducted, 'a change in practice' from revolution conceptualised as war (Robinson, 1996b, 451). As its creators Vlad Taneev and Marina Tokareva explain, eco-economics involves consideration of issues of carrying capacity, co-existence, counter-adaptation, legitimacy mechanisms and ecologic efficiency, among other ecologically oriented issues, along with the recognition of the co-existence of cultural and natural domains: the practice of economics on Earth is described as a 'deformed offshoot' of ecology (Robinson, 1996c, 351). It is a 'synthesis of systems' 'based [as Vlad explains] on models from Terran history, and its various parts have all been tested on both worlds, and have succeeded very well' (Robinson, 1996a, 148). Many of these economic systems have been exapted from real-world systems currently practised on Earth. The microeconomics of eco-economics is borrowed from the 'Mondragon region of Spain[, while the] different parts of the macroeconomy have been used in the pseudo-metanat Praxis [(a fictional corporation)], in Switzerland, in India's state of Kerala, in Bhutan, in Bologna Italy, and in many other places, including the Martian underground itself' (148). Eco-economics is supplemented by a barter and gift system, the former of which is based on a 'hydrogen peroxide economy, where things are priced by calculations of their caloric value,' the latter 'a nitrogen standard' covering 'two planes, the need and the gift' (Robinson, 1996b, 463).

This system ties the economic practices of the Martian government directly to the life-support system of the planet. They are slowly pieced together from a variety of economic systems from Earth in a manner that resonates with Rasula's notion of 'wreading,' 'the collaborative momentum initiated by certain texts, like the *Maximus Poems*, in which the reader is enlisted as an agent of the writing,' or a 'nosing into the compost library' (2002, 11, 18). Vlad and Marina explore the compost library of economic systems, carefully selecting examples for exaptation into a new Martian economy that brings the ecological aspects of human dependency on nature to the fore. Others are also involved in developing and testing components of this system in a dialogic process that leaves the act of creation open. To Boone eco-economics sounds like 'echo economics,' an ambiguity that emphasises the exaptation of practices from Earth's compost library: echoes of Terran history and storytelling that are given new life in an alien ecological system (Robinson, 1996c, 351).

There is a danger that using the Martian landscape as a space to compile a new culture from Earth's compost library will overwhelm Mars's alien otherness and undermine its own status as an independent and autonomous nature. Clayborne warns that the result of terraforming will be that 'Mars will be gone and we'll be here, and we'll wonder why we feel so empty. Why when we look at the land we can never see anything but our own faces' (Robinson, 1996c, 190). In this allusion to the final story of *The Martian Chronicles* – in which a family identifies themselves as Martian after seeing their reflections in a stream (Bradbury, 1958, 222) – Robinson highlights how the failure to recognise Mars's otherness results in an alienating emptiness that undermines the possibility of reaching an identifying engagement with the land. In contrast, Clayborne sees the planet as a space with its own meaning: '[t]o see the landscape in its history, to read it like a text, written by its own long past; that was Ann's vision, achieved by a century's close observation and study, and by her own native gift, her love for it' (Robinson, 1996a, 98). Its geologic and climatic processes, while not part of a life-support system of its own, leave traces of a 'voice' that can be read with the appropriate scientific knowledge. The Martian landscape has another history, 'the history of Mars in the human mind,' or the Martian megatext, that compost library constructed in part by science and sf and in part by older forms of knowledge about Mars (Robinson, 1996c, 13). The Martian landscape is a palimpsest written upon by physical and intellectual landscaping processes; alternatively, it is, from the perspective of some colonists, a tabula rasa without a history, a landscape of 'immense potential [... a] blank red slate' (108). The struggle over Mars's meaning is the main issue at stake in the *Mars* trilogy, and it subsumes the narrative of revolution that constitutes much of its political engagement.

Nirgal, one of the first generation of Mars-born colonists, gives a speech on the occasion of his visit to Earth in which he suggests that 'Mars is a mirror [...] in which Terra sees its own essence.' Characterising Mars as an expression of 'Terran thought and Terran genes,' Nirgal sees the purifying voyage to the planet as an opportunity for the colonists to 'help the home planet by serving as a way for you to see yourselves. As a way to map out an unimaginable immensity' (Robinson, 1996a, 178). Mars offers a space in which a new composting library of practices and institutions can be explored, ideas that have been developed through the colonists' various relationships to Mars. Their physical relationship to the new planet prompts the development of new socio-economic relationships and new philosophies to meet the requirements of habitation. 'As people learn more, [says Nirgal] they understand better their dependence on each other and on their world. On Mars we have seen that the best way to express this interdependence is to live for giving, in a culture of compassion' (178). This view works as a counter to the interests of Earth, whose own politico-economic structures, dominant on Mars throughout the narrative of *Green Mars*, are overthrown to make way for the sifting and exaptation that goes into creating a new human relationship to the planet during *Blue Mars*.

Thomas J. Morrissey accounts for the relationship between Mars and Earth in many of the stories of the Martian megatext as one in which they are 'bound like jealous siblings or inconstant lovers, alternately brought together or torn asunder by intelligent but often conflicting visions, often expressed in metaphor' (2000, 372). Earth's ecological failures form a background and foil for economic developments on Mars, and in the *Mars* trilogy eco-economics offers a challenge to Earth's own economic systems. Earth's practices are described as a 'cycle of madness,' a life-support system that is detrimental to continued habitation (Robinson, 1996b, 637). One character says of Earth that '[w]e have been liquidating our natural capital as if it were disposable income, and are nearing depletion of certain capital stocks, like oil, wood, soil, metals, fresh water, fish, and animals. This makes continued economic expansion difficult' (100–01). That many of the multinationals who invest in Mars intend for the planet to become a field for further capitalist expansion, given the dwindling capital and field for growth on Earth, is testified to by their movement of corporate security forces onto the planet and their aggressive and intrusive interference in the lives of the workforce that they ship to Mars. Earth's governments see the red planet as a site for the relief of the population surplus and as an answer to ecological crisis and conflict on Earth. One character notes that 'carrying capacity was a very fuzzy abstract concept, depending on an entire recombinant host of complexities such as soil biochemistry, ecology, human culture' (Robinson, 1996a, 346). Carrying capacity is thus an ecological principle that illustrates the interdependency of physical and cultural parameters in determining the appropriate level of strain that a life-support system can bear. Earth and the multinationals push for increased immigration quotas in order to solve the problem of overpopulation and to create a new market for economic expansion on Mars. Morrissey summarises these values and practices as part of what he calls a 'Dominant Social Paradigm', which received widespread representation in the stories of the Martian megatext prior to the late 1980s–1990s, after which many texts move towards what he characterises as the 'New Environmental Paradigm,' of which Robinson's own *Mars* trilogy is exemplary (2000, 386).

Against this background, the Martian revolution aims to establish new expectations for habitation and economic practice based on scientific, ecological principles: '[s]cience is creation,' argues Sax (Robinson, 1996c, 213). As the narrator notes, '[m]etanational capitalism's track record at this point did little to support it; in the last century it had precipitated a massive war, chewed up the Earth, and torn its societies apart. Why should they not try something new, given that record?' (Robinson, 1996a, 148). Examples of this movement away from the economic systems of the past include the pseudo-metanational corporation Praxis. This corporation aims to develop new possibilities for economic relationships on Earth, and it allies itself to the Martian revolutionaries in order to learn from the social experiments taking place there during and after the revolution.

Praxis establishes new industries on Mars that engage in ecologically oriented industrial practices: a local Praxis salvage subsidiary fittingly named Ouroboros provides an example of an economic endeavour tuned to the necessities of maintaining a life-support system on Mars: 'there was not a large garbage output on Mars; almost everything was recycled or put to use in creating agricultural soil, so each settlement's dump was really more of a holding facility for miscellaneous materials, awaiting their particular reuse' (Robinson, 1996b, 131). Ouroboros 'transforms waste into resource' through 'green' nanotechnology, which Colin Milburn argues is 'the symbol of corporate domination' and which, in this example, 'is remade and remobilized as "power from below"' (2012, 73). This mobilisation of power can thus be seen as a metaphorical exaptation. Milburn argues that 'Robinson shows us that science fiction is itself an instrument of environmental nanopolitics, a molecular technology for terraforming our world and ourselves' (57). Terraforming and the mythic image of Ouroboros offer metaphors for an sf composting library that, through the exaptation of elements that establish a New Environmental Paradigm, is oriented towards a future that seeks to distribute political power.

Morrissey notes that Robinson's *Mars* trilogy is engaged in a 'search for a vision that can sustain us in the future' (2000, 386). Mars, as Nirgal suggests, cannot save Earth by functioning as a safety valve for immigrants wishing to escape from the Dominant Social Paradigm of Earth, but it can function as a way to revitalise Earth's socio-economic and political institutions through the developments generated by the Martian compost library. The two planets have always been connected, despite the claims of some of the hardliners amongst the revolutionaries. There are two dimensions to the challenge of this reconnection: the relationship between Earth and Mars must be redefined, as Nirgal attempts to do, and the Red/Green debate on Mars must be resolved. This debate pivots on the disagreement over terraforming, with Reds supporting a preservationist stance towards Mars's natural otherness and Greens emphasising the transformative potential that life offers to the colonists. The narrative moves towards a synthesis between opposed positions as much as it attempts to resolve the conflict between Mars and Earth. Sax, initially the strongest proponent of a heavy industrial terraforming model, finds that his preferences for the Martian planet are transforming as he discovers Mars's own voice expressing itself through the new life being introduced to its surface: '[f]arther on lay some tangles, red-stalked, greenneedled, like beached seaweed in miniature. Again that intermixture of red and green, right there in nature staring at him' (Robinson, 1996a, 67). As Earth's compost library is sifted through by the colonists, the Martian landscape offers metaphors for a synthesis between opposed positions, offering a symbolic point of reconnection between the compost libraries of the two planets.

As the trilogy progresses these themes are argued over and considered from an increasing number of different perspectives and contexts.

The *Mars* trilogy foregrounds the structural relationship between the chronotopes of Earth and Mars, underscoring a postcolonial dimension to terraforming other planets, aspects of which have already been discussed in terms of national and global identity and familiar and unfamiliar spaces. Edward Said explains that 'there is no doubt that imaginative geography and history help the mind to intensify its own sense of itself by dramatizing the distance and difference between what is close to it and what is far away' (2003, 55). The vacuum of space, a chronotope symbolising a purifying transformation, signifies the radical distance between the two planets and serves as one of several estranging devices. This distance, and the imaginative geography already associated with Mars, establishes it as a field for the exploration of and experimentation in alternative social and individual identities that allow Earth's population to see distorted reflections of themselves modified by influences from the alien landscape, but only if it is not landscaped solely in terms of the chronotope of the interplanetary mine.

When applied to sf worlds, Said's notion of Orientalism sheds light on Clayborne's resistance to the terraforming of Mars, which is geared to recreating the self, symbolised by Earth's landscape, on another planet. Clayborne's and Cherneshevsky's response to the wilderness of the Martian landscape is an affective response to nature's otherness, a non-human identity that remains outside of the familiar bounded spaces of the colonising outposts. Terraforming reduces these spaces to a partial identity through the growth of an ecological system imported from Earth: an ecological imperialism. This identity is partial because such imports must be adapted to the alien environment and because unadapted imports are subject to the evolutionary influence of these alien spaces. The social aspect of settling other planets and the ecologically focused terraforming project are not simply two material necessities to ensure survival on Mars, but attempts to reduce Mars's otherness to an identity. This notion is encapsulated by the phrase 'terraforming' itself: the colonists attempt to change the Martian landscape, which is other from the point of view of Earth's population, into another Earth.

Nirgal's description of Mars '[a]s a way to map out an unimaginable immensity' elides the trilogy's concern with the notion that the Martian landscape exists independently of Earth's interests (Robinson, 1996a, 178). Nevertheless, he points to the importance of landscape as 'hero' or 'character' in sf, mobilising language that taps into sf's sense of wonder, a character's and (potentially) a reader's response to a conceptual breakthrough, a shift in the conceptual paradigms framing views of the world. Mars's function as a way to map the future is one such conceptual shift that connects Earth and Mars. The trilogy explores an imagined future to address human-centred concerns that cannot be predicted from our present vantage.

Said discusses the example of 'Bouvard's vision of Europe regenerated by Asia,' which 'represents what Flaubert felt to be the nineteenth-century predilection for the rebuilding of the world according to an imaginative vision, sometimes accompanied by a special scientific technique' (2003, 114). He notes that '[k]nowledge of the Orient, because generated out of strength, in a sense creates the Orient, the Oriental, and his world' (40). In this sense the reconstructed other is an identity, the result of projection that excludes the other from opportunities to define itself. Nature has no obvious voice and so is at greater risk of being spoken for and its otherness being elided. Denial of otherness provides foundations for the repetition of the historical influences that Bogdanov advocates against in breaking away from history. This apparent mirror is mediated by the effects of the Martian environment on the colonists, who do respond to the otherness of Mars and adapt their culture to account for new forms of local and global belonging. Through the synthesis of self and other Nirgal offers Earth an opportunity to regenerate itself in a manner that resonates with Said's examination of Bouvard's (sometimes scientific) utopian vision.

The *Mars* trilogy constructs textual spaces populated by competing ideological positions. The language of sf – its use of megatextual images such as chronotopes – maps dialogues between positions for examination and interrogation. The spaces of Mars and Earth, and the colonising outposts and mining facilities, take part in a heteroglossic dialogue in which different voices are placed in relation to each other to create a polyphonic whole that offers an examination of societies in relation to their environment. The *Mars* trilogy shows how the meaning of these chronotopes is contested. The Martian landscape is spoken for from the position of transnationals, Greens engaging with the terraforming process, and Reds agitating for the preservation of Mars as wilderness. These debates embody the process of world-building on the social level, where a new Martian identity, distinct from Earth, arises from the interstices of debate and from the Martian landscape's influence. This world-building engages with eco-cosmopolitan concerns and is linked to utopia and dystopia, two alternative paths down which the development of the new society could lead.

Science and Nature

Science offers sf a series of discourses that terraforming stories adopt and adapt to inform literary constructions of alien landscapes. The initial motivation to transform alien planets into new Earths puts into play contesting interpretations of ideal relationships to the environment. Scientific discourse dominates these 'discussions,' accruing metaphorical implications (often socio-political) and offering alternative perspectives on the environment that variously undercut or contribute to other modes of understanding. Robinson connects science to socialism, distinguishing it from nineteenth-century scientism by explaining that 'the "scientific" returns as a way of talking about providing some kind of ecological basis to economy [...] Another way it returns is to regard science itself as a utopian project and as a form of human interaction' (see Buhle, 2002, 89). Through the course of such narratives, as colonists struggle to re-territorialise their new environments, the feedback between science and non-scientific knowledge generates alternative conceptions of the nature and import of alien spaces, forcing us to rethink the concept of nature and of human relationships within these environments.

Because of his early efforts as head of the terraforming project in pioneering a heavy industrial terraforming model, Russell is closely aligned with the values linked to applied science as an expression of colonial ideological value. He initially behaves like a caricature of the scientist figure who prefers to separate politics and science. Speaking of instances in which, as with the cell–culture metaphor in Sargent's trilogy, nature is used to account for social relations, Russell argues, 'I don't think it helps to make analogies between the physical and social worlds' (Robinson, 1996c, 646). Russell's fundamental disagreement with Clayborne over the future of Mars centres around Russell's failure, in Clayborne's view, to 'see' the landscape: his early perception is mediated by a range of scientific apparatus and audio-visual technologies, and he defines Mars's value in terms of its value to the terraforming project. Clayborne's view is informed by her scientific awareness of 'deep' geological time and by her experience of the landscape. This opposition reflects Rolston's participatory versus objective view of the environment, with Clayborne's preferred method of relating to Mars involving long periods of physical exploration of, participation with and perception of Mars. Yet her experience is combined with a scientific view that shapes her awareness of global processes. Just as science can, as Rolston argues, lead to a better appreciation of the environment, something more may be needed, perhaps something '[t]hat may go beyond science, but [which] must go through science to go beyond' (1995, 375).

The problem of developing a sense of place that includes an aesthetic appreciation of the environment is a perceptual issue, and science remains a powerful tool for developing these perspectives. This is joined to a political problem, one of persuading others to view the landscape in like manner. In *Red Mars*, Clayborne reflects that her inability to understand Russell and vice versa is due to '[v]alue systems based on entirely different assumptions. Completely different kinds of science' (Robinson, 1996c, 649). As Russell struggles to understand and reconcile his longstanding disagreement with Clayborne in *Green Mars*, he suggests that their difference centres on 'the fact-value problem. Science concerns itself with facts, and with theories that turn facts into examples. Values are another kind of system, a human construct' (Robinson, 1996b, 185). Despite the problems Clayborne raises in response, that science is also a human construct and that it too has values embedded within its scientific method, Russell insists that terraforming as applied science involves a choice of how to utilise the insights gained from science, and so is a value problem, whereas science and the scientific method itself are concerned only with facts.

By *Blue Mars* Russell modifies this view to articulate a vision of science as utopian and objective, 'a social construct, but [...] also and most importantly its own space, conforming to reality only; that was its beauty' (Robinson, 1996a, 677). He describes how science is built on a method that privileges dialogue and adaptation, arguing that '[i]n truth the work of science was a communal thing, extending back even beyond the birth of modern science, back all the way into prehistory, as Michel had insisted; a constant struggle to understand' (676). Russell remains convinced that the communal practice of scientific enquiry in this general sense is the source of as much of the value of science as the facts and theories that it generates. He experiences an epiphany during a moment when he 'steps back' to reflect on the practice of a 'Martian' science that indexes the growth of a new global sense of belonging to the alien planet:

> something inside him would glow till it hurt, some parasympathetic reaction spilling out of his limbic system – now this was science, by God, this was Martian science, in the hands of the scientists themselves, working together for some collective goal that made sense, that was for the common good; pushing at the edge of what they knew, theory and experiment bouncing back and forth like a blur of Ping-Pong balls, week after week finding out more, going after more, extending the great invisible parthenon right out into the uncharted territory of the human mind, into life itself. It made

> him so happy that he almost didn't care if they ever figured things out; the search was all. (Robinson, 1996a, 702)

This Baconian image of the utopian impulse in science relocates the focus of the scientific utopia from the construction of societal blueprints to one emphasising a dialogic process whereby scientific engagement with the natural world ramifies through social worlds. This view of the practice of science as a communal activity accords importance to the means by which knowledge of the natural world is generated. The *Mars* trilogy scrutinises the structure of the scientific enclave and its communal goal, which, taken as a model, raises the question of its connection to the wider political sphere. In *Red Mars*, Bogdanov tells Boone that 'a scientific research station is actually a little model of prehistoric utopia, carved out of the transnational money economy by clever primates who want to live well,' and that their unwillingness to tackle the problem of 'work[ing] to create such conditions for everyone' compromises its status as a genuine utopia (Robinson, 1996c, 402, 403).

Russell's outlook towards science changes in response to social upheaval. As the head of the terraforming project and in his role as one of the mythic 'generals' of the Martian revolution for independence from Earth, Russell is positioned at the forefront of a political struggle for defining Mars, forcing him to grapple with the problem of the fact–value interface, of working to establish a foundation for the practice of science, and of developing a genuinely communal decision-making process. Suffering from aphasia after capture and torture in *Green Mars*, Russell slowly recovers his linguistic ability alongside a new awareness of social interchange. He begins to consider the importance of other modes of understanding in *Blue Mars*: '[s]ymbolic value: it was a concept Sax was trying very hard to understand. [...] symbol, "something that stands for something else," from the Latin symbolum, adopted from a Greek word meaning "throw together." Exactly. It was alien to his understanding, this throwing together, a thing emotional and even unreal, and yet vitally important' (Robinson, 1996a, 47). Russell's reflection on meaning takes him into the figurative realm and leads him to combine his understanding of science with this notion of symbolic meaning, not to blur their boundaries, but to complement his experience with another perspective. He makes connections between the 'throwing together' of the symbol, the mind, ecology, and climatology, and uses metaphors drawn from the science underlying terraforming to suggest models that characterise the mind itself: 'an ecology – a fell-field – or else a jungle, populated by all manner of strange beasts,' or 'chemical energies surging hither and yon, like weather in an atmosphere'

(Robinson, 1996a, 55): '[t]hat was better – weather – storm fronts of thought, high-pressure zones, low-pressure cells, hurricanes – the jet streams of biological desires, always making their swift powerful rounds ... life in the wind. Well. Throwing together. In fact the mind was poorly understood' (Robinson, 1996a, 55). This metaphor for the human mind utilises the chaos-influenced science of meteorology in a manner that echoes Sargent's spatialisation of Venus's hostile environment and the domes, all of which operate as metaphors for the mind turned inward. Ecology and climatology are figuratively aligned to the indeterminacy involved in analyses of the otherness of the human mind. Ecology takes on a figurative status as the science of 'throwing together' as it depends on a range of scientific and cultural disciplines to portray and grapple with the entirety of an ecological network that includes humanity and its adaptations to multiple environments. Turner suggests that sexual reproduction recapitulates this throwing together and that '[o]n rare occasions that "symbolon" can be the start of the great grand poem of a new species.' In his view, art and language are extensions of biological and ecological processes of throwing together: '[p]oetry is fast evolution: evolution is slow poetry' (see Pak, 2014b, 7).

Pamela Sargent, Frederick Turner and Robinson imagine several levels of engagement with other worlds that form a complex of interacting or contesting discourses of varying degrees of dominance. These works establish a plurality of discourse, a model of overlapping interests that intersect and diverge, but are ultimately brought together in the physical space of the terraformed world. Sf has offered an especially powerful language for expressing ecocritical concerns, but this has risen slowly to the foreground in its own distinctive ways in terraforming stories. The influences feeding into or implied as narrative possibilities are used to explore counter-models to inadequate relations between the human and non-human.

Rolston ends his paper by reflecting on various prescientific views of the environment, arguing that '[s]cience should demythologise these views but must itself find a new myth that encourages appropriate aesthetic responses to nature' (1995, 384). He is claiming not that traditional cultural responses to nature are detrimental or irrelevant, but that scientific understanding necessarily needs to demythologise them in order to be science. Just as Clayborne argues in *Red Mars*, Rolston claims that 'humans are always the landscape architects, and even science is another cultural way of framing landscape' (376). Terraforming, as a literature of landscaping, partakes in the throwing together theorised by Russell, combining scientific, political and cultural parameters into the space of its thought experiment. It uses the dialogism inherent in scientific enquiry that Russell identifies to construct images of science and society that offer alternatives to imperial approaches to nature. The science of the *Venus* trilogy also opens up reflection on the use of projects of applied science to enforce repressive socio-cultural and political agendas. Terraforming narratives explore science's limits and investigate ways in which prescientific and scientific modes of understanding fuse. They concern themselves with the struggle to transform and use the sciences, especially ecology, to construct new myths for conceptualising and relating to nature and society.

On Martian Myths

Commenting on Bud Foote's description of the 'self-conscious intertextuality' of *Red Mars* as drawing attention to the novel as an artefact that encompasses older stories, Franko notes that 'Mars itself is the nexus of many of these embedded stories, from science fictions to fictional canals to ancient myths of Mars inspired by its redness and erratic revolution' (1997, 59). Mars is the iconic basis of the parabolas explored in the trilogy, with terraforming functioning as a second-order cluster of icons that modify the trajectory of the parabola in various ways (through, for example, the motif of soil and compost). Several characteristically Martian myths dominate the imagination of the colonists, all of which are interlinked and build upon the compost library of Earthbound myth and science in the new context of the Martian landscape:

> stories have naturally blossomed to fill the gap, just as in Lowell's time, or in Homer's, or in the caves or on the savannah – stories of microfossils wrecked by our bioorganisms, of ruins found in dust storms and then lost forever, of Big Man and all his adventures, of the elusive little red people, always glimpsed out of the corner of the eye. And all of these tales are told in an attempt to give Mars life, or to bring it to life. Because we are still those animals who survived the Ice Age, and looked up at the night sky in wonder, and told stories. (Robinson, 1996c, 14)

The little red people of Mars, often seen in the corner of the eye but never directly, is an exaptation of sf tales of 'little green men' and tells of an indigenous people who adapt to the influx of colonists and their ecologies. As ants are introduced as part of the project of soil construction, a mythic story of this event arises that describes the little red people's encounter with these creatures: they 'were just the right size to ride, it was like the Native Americans meeting the horse. Tame the things and they would run wild' (Robinson, 1996a, 113). The ability of the little red people to adapt to the colonists is testified to in tale after tale until in *Green Mars* some of the colonists begin identifying themselves with the myth: one character reflects that '[t]hey were ants in such a landscape, they were the little red people themselves' (Robinson, 1996b, 326), while at a conference to outline a governmental system for Mars, an anonymous individual writes the slogan 'However: We Are the Little Red People' on a public message board (Robinson, 1996a, 156). The official Martian constitution operates, Burling argues, as a referential framework, 'a provisional set of shared beliefs' that makes a radical democracy on Mars possible (2005, 80). The myth of the colonists who become Martians complements this official political document with another referential framework, a mythic origin story of transformation from a colonial, capitalist annexe of Earth to an environmentally transformative society that shapes its values around the demands of living on Mars so as to extend ethical considerations to Mars's non-human nature.

The classic pioneer myth of Paul Bunyan and his big blue ox Babe is also subject to exaptation on Mars. Bunyan and Babe are characters who feature in several traditional pioneer tales, one of which describes how Bunyan finds and adopts the calf Babe during a winter's day, amidst drifts of blue snow. Complementing Bunyan's own stature, Babe grows to gigantic proportions after returning to Bunyan's camp. The most detailed appropriation of these figures in the *Mars* trilogy relates to Bunyan's encounter with Big Man from big planet, with whom he engages in a contest of strength. In an echo of some Australian Aboriginal myths, their contest transforms the landscape of Mars by creating many of the named geological features, from Argyre and Hellas to Nirgal Vallis, Ceraunius Tholus and the Elysium Massif. The contest kills Bunyan, '[b]ut his own bacteria ate him, naturally, and they crawled all around down on the bedrock and under the megaregolith, down there going everywhere, sucking up the mantle heat, and eating the sulfides, and melting down the permafrost. And everywhere they went down there, every one of those little bacteria said I am Paul Bunyan' (Robinson, 1996c, 454). This passage recalls Alfred Bester's short story 'Adam and No Eve' (1941), which tells of an engineer's attempt to reach space in a rocket that uses a gas and a combustion process that ultimately destroy all life on Earth. The sole survivor – the inventor – drags himself into the empty sea so that the micro-organisms in his body can repopulate the Earth. Bunyan, a representative of America and its pioneer past, is bested by something even bigger than him, Big Man, who stands for the vastness of the wider solar system. Their struggle is a mythic retelling of the struggle of the colonists on Mars, while the death of Bunyan symbolises the death of America, and indeed nations, as the dominant players with interests in the interplanetary colony – a role that the transnationals in the later parts of the trilogy assume, and which later still is superseded by the rise of the Martian government. Bunyan's transformation into bacteria mirrors the human colonisation of Mars; just as the bacteria colonise Mars, the colonists transform the Martian landscape through ecopoietic means. The metaphor embedded in this tale reaches towards a vision of consensus in which the Martian landscape is changed through the combined efforts of a multitude. Little red people or the bacteria of Bunyan – these tales are structured as parabolic arcs that embody the exaptation of stories into myths that function as metaphors for the creation of a new society embedded in its landscape. They rework elements from the compost library of Earth within the context of a terraformed Mars.

Bardini's notion of junk and its tendency to resist closure and Rasula's notion of the compost library offer ways to consider what is characteristic of the sf motif of terraforming. Centred on the creation of life-support systems on other planets, terraforming Mars depends on the initial modification of the atmosphere, but ultimately on the construction of soil, or compost. These physical parameters form the essential basis of a Martian eco-economics that ties both the physical and the socio-economic aspects of the colony into a whole. The Martian landscape is threatened not only physically but intellectually, in the sense that the projection of human interests onto the planet poses the risk that its nature will be overlooked in favour of using its landscape solely as a field for autological speculation and a recycling of Earth's practices, which threaten to close possibilities for creating new, more ecologically sound, modes of habitation. Nevertheless, when the eco-economic system is tuned to the specificities of the Martian landscape, the potential for feedback from the alien to the human offers to transform the multiple voices embedded in Earthbound history into new avenues for socio-cultural experimentation. Burling notes that this experimentation is provisional and possible only through struggle, requiring the continual revision of previously stable points of social and political agreement in contexts that are subject to change (2005, 83). The challenges to Earth's socio-economic systems posed by the innovations in social thought developed on Mars offer in turn to revitalise the compost library of Earth.

Rasula states that '[a]mong available modes of discourse, poetry is unique in favoring utopia as transient occasion, not universal city. Poems effectively consume all the energy they generate' (2002, 71). Robinson's use of terraforming to explore the juncture between ecology, politics and society favours transience of another sort, a utopia of process. Robinson explains in an interview that 'I will always remain a science fiction writer because we live in a giant collaborative science fiction novel that we are all writing together. It is the realism of our time, especially in the industrial West, but more and more everywhere' (see Buhle, 2002, 90). This view of sf as an integral part of a wider dialogue that includes economic, scientific and technological knowledge and its impact on socio-political practices, lifestyles and thought positions the sf compost library as an indispensable ecological literature. Robinson's focus on the impact of science and technology on society offers the reader a way of thinking about sustainable ecological processes, the extent to which they can be modified, and the possible outcome of these modifications. Attentiveness to the sf compost library is also an attempt to engage with the world outside of the sf megatext through a collaborative writing process that takes as its basis a distrust of static utopias, favouring instead the openness of utopia as a continual process.

The parabolas of Robinson's terraforming narrative connect scientific ideas and their practical implications to metaphors for social and political philosophies that model an ecological approach to habitation, thus working in environmentally transformative ways to critique the limits of contemporary society's economic, political and social institutions. Investment in dynamic utopias finds in the motif of terraforming as a life-support system a figure that embodies ideas of exaptation, junk, cybernetics, and open feedback mechanisms, and is emblematic of the psychic and social interventions of human communities upon their environments. The *Mars* trilogy engages in ideas of bootstrapping as an iterative, co-adaptive learning process, not just for the colonists of the text, but for contemporary (w)readers of sf, whose vacillation between literal and metaphorical readings of the narrative creates a feedback loop that generates new perspectives on nature and society that cannot be reduced to the initial sf motifs underpinning the narrative. The exaptation of the sf compost library is thus a process that, like Attebery's parabolas of sf, generates new and creative ways of exploring ecologically oriented modes of habitation.

Conclusion

Narratives in the terraforming tradition contribute to a polyphony of worldviews about society and its ideal relationship to its environment and to nature. Landscaping processes act as interfaces for this preoccupation with society and with the relationship between the human and non-human. The pastoral, utopian discourse, ecology and environmentalism, along with elements of the sf megatext, have fed into the structure of terraforming narratives and shaped their thematic engagements. The megatext that the terraforming tradition constructs is Bakhtinian in its propensity to interrelate multiple texts and voices within the textual space of the terraformed world. The process of world-building that the terraforming narrative enacts provides a forum for philosophical, socio-cultural and political enquiry that is ecological in character. Drawing from the insights of James Lovelock's Gaia hypothesis since the 1980s, terraforming has become a primary motif for environmental and ecological thought in sf discourse. At the level of discourse, these stories re-use narrative elements and tropes developed by earlier texts, sometimes recycling these elements but also transforming them into new engagements with contemporary social and ecological anxieties.

This book has shown how the development of the idea of terraforming first arose within sf discourse before influencing scientific speculation about the future and environmental philosophical thought during a period of anxiety about climate change. This influence generated a feedback loop between sf, science and environmental philosophy, which helped shape the development of a narrative that could provide fertile ground for engagement with contemporary environmental concerns. Chapter one demonstrates how the first terraforming and proto-Gaian narratives can be read against environmental philosophical themes such as landscaping and establishes the relationship between nature's otherness and the representation of alien environments and organisms in sf. Chapter two analyses the influence of the American pastoral in the terraforming narratives of the 1950s boom and draws connections between landscaping processes, Bakhtin's chronotope and the sf megatext. Chapter three explores the influence of the burgeoning environmental movement on the development of representations of terraforming in the period 1960–1970, while chapter four considers how terraforming was consolidated and grew more sophisticated in its representation in the works of the 1980s. The last chapter is focused on an analysis of Kim Stanley Robinson's acclaimed *Mars* trilogy and the major trends in the development of the terraforming narrative in the 1990s. To conclude, this section considers some of the terraforming narratives that were published after Robinson's *Green Mars* in order to demonstrate how the motif continues to offer resources for ecological and socio-political thought.

Terraforming narratives are preoccupied with the problem of creating a new human history that can escape, resolve or transcend the failures of the past. Perhaps inspired by the success of Robinson's *Mars* trilogy, but certainly by the increasing global attention that climate change received, many terraforming stories were published during the period of the 1990s–2000s. Brian Aldiss and Roger Penrose's *White Mars* (1999) engages in dialogue with Robinson's *Mars* trilogy. Drawing on the utopianism of *Ecotopia* (1978 [1975]) and *The Greening of Mars* (1984), *White Mars* advocates a non-interventionist model of engagement with the solar system that is grounded in the intrinsic value of non-human nature. Like Allaby and Lovelock's *Greening*, *White Mars* is a collaboration between an sf writer and a scientist, thus underlining the importance of science as an essential element informing the sf imagination of landscape. Although an important text because of its explicit engagement with others in the terraforming tradition and for its philosophic and scientific debate over issues of environmental ethics, it has had comparatively little impact on other terraforming narratives.

Jack Williamson returned to terraforming with five short stories and novelettes comprising the novel *Terraforming Earth* (2003 [2001]), which continues the trend towards reflection on the future planetary changes that the planet might undergo. Williamson's narrators are clones created by an automated process at a lunar base intended as a sanctuary in case of catastrophe on Earth. These narrators allow Williamson to explore changes to Earth in several distinct periods of time, each separated at the geological scale. Echoing his earlier short story 'Collision Orbit' (2004 [1942]) – in which 'terraform' was first coined – an impactor threatens Earth before the base can be completely prepared and a suitable staff organised. An ad hoc group is hastily assembled to ensure that genetic material and appropriate information can be supplied to the future clones, who slowly repopulate the Earth and who stand as witnesses to a future history inhabited by strange creatures and colonising aliens. The overall thrust of these linked stories emphasises the plasticity of Earth's topography while illustrating the continuity and differences that underlie human societies. The stories explore the conflicts that alien and human groups continually struggle with over the course of their histories. Like Earth's topography, these civilisations are malleable and demonstrate a wide range of variation.

A group of short stories and novelettes published at this time further illustrates the tendency for the terraforming narratives of this period to turn towards a consideration of Earth and its history. Joe Haldeman's 'For White Hill' (2001 [1995]) is set on a far-future Earth devastated in the 'Extermination,' during which the alien Fwndyri unleashed their nanophages, intelligent nanomachines programmed to deconstruct human DNA and later reprogrammed to sterilise all life on Earth. A Council of Worlds headquartered on Earth is formed in response and several people, the narrator and the eponymous White Hill, for instance, are '"bred" for immunity to the nanophages' (Haldeman, 2001, 246). The story centres on a competition to commemorate the conflict and devastation of Earth, the competitors of which are selected artists from around the galaxy: '[b]efore they reterraformed the Earth, though, they wanted to isolate an area, a "park of memory," as a reminder of the Sterilization and these centuries of waste, and brought artists from every world to interpret, inside the park, what they had seen here' (Haldeman, 2001, 250–51). Among other monuments of Earth's history, the artists tour the Grand Canyon and Chicago, the Pyramids of Giza and Rome for inspiration. During the tour they receive word that the competition has been profoundly changed: '[s]omehow the Fwndyri have found a way to make its [the sun's] luminosity increase' (Haldeman, 2001, 260): they have found a way to accelerate the aging process of the sun, thus resuming hostilities with Earth.

Memory is central to this story. The narrator begins by stating, 'I am writing this memoir in the language of England, an ancient land of Earth, whose tales and songs White Hill valued' (Haldeman, 2001, 245). White Hill tells the narrator that she did not tour the domes for inspiration for her piece because '[a]ll the story's here, anywhere. It isn't really about history or culture' (247). Petrosian, White Hill's native language, possesses a palindromic mood (an inflection of verbs that facilitates the structuring of sentences as palindromes) that allows statements such as '[d]reams feed art and art feeds dreams' to be conveyed by three words – although translating these statements into English makes their meaning uncertain (247). This awareness of language, of the memories of monuments and of the land itself, is woven into White Hill's philosophy:

> The reality is that it is all one to them. What makes Seldenians so alien is that their need for balance in life dissolves hierarchy: this piece of art is valuable, and so is this orgasm, and so is this crumb of bread. The bread crumb connects to the artwork through the artist's metabolism, which connects to orgasm. Then through a fluid and automatic mixture of logic, metaphor, and rhetoric, the bread crumb links to soil, sunlight, nuclear fusion, the beginning and end of the universe. Any intelligent person can map out chains like that, but to White Hill it was automatic, drilled into her with her first nouns and verbs: Everything is important. Nothing matters. Change the world but stay relaxed. (Haldeman, 2001, 266)

This ecological vision offers an alternative way of ordering the world. Art, sex and food – all three retain their distinctions, but their value is levelled as the relationships between them and other entities are foregrounded. The narrator continues to reflect on the English language throughout the narrative, noting that 'England's versatile language, like mine and hers, is strangely hobbled by having the one word, love, stand for such a multiplicity of feelings' and that '[p]erhaps that lack reveals a truth, that no one love is like any other' (Haldeman, 2001, 265). Experience is larger than the capacity for language to record, and memory suffers from the problem of satisfactory representation. The narrator's memoir, and the theme of art itself, offers a potential avenue for representing this ecological vision that the reader is implicitly asked to judge: does his record do justice to White Hill's memory and the narrator's emotional investment in Earth over the course of the events of the story?

The problem of creating artistic monuments as a way to commemorate history is captured by the aesthetic division between the artists after they receive news of the resumption of conflict with the Fwndyri:

> We had divided into two groups, and jokingly but seriously referred to one another as 'originalists' and 'realists.' We originalists were continuing our projects on the basis of the charter's rules: a memorial to the tragedy and its aftermath, a stark sterile reminder in the midst of life. The realists took into account new developments, including the fact that there would probably never be any 'midst of life' and, possibly, no audience, after thirty years. (Haldeman, 2001, 267)

White Hill claims both sides; the narrator is an originalist. As an empathic therapist and practitioner of 'jaturnary,' White Hill is able to share in and to connect other conscious minds (Haldeman, 2001, 252). She decides to submerge her ego to join the thousand inhabitants of Earth in a potentially thousand-year cryogenic sleep aboard a spaceship that will allow them to escape the devastation of the planet. The narrator's memoir is a monument raised to her name. It is to be translated into as close a universal language as Earth's xenobiologists can contrive, cast in a material that will outlast the centuries and appended to a document recording the sum of Terran knowledge: 'a standard book that starts out with basic mathematical principles, in dots and squares and triangles, and moves from that into physics, chemistry, biology. Can you go from biology to the human heart? I have to hope so. If this is read by alien eyes, long after the last human breath is stilled, I hope it's not utter gibberish' (Haldeman, 2001, 275). This uncertainty emphasises the transience of language and memory and casts a critical eye over the monuments that are experienced during the tour. Earth in its present state is one such monument, prompting White Hill to exclaim that '[t]his is too big and terrible a thing. I feel like an interloper. They've lived through centuries of this, and now they want us to explain it to them?' (Haldeman, 2001, 254).

From an environmental philosophical perspective, this story is a *tour de force* of landscaping. The way humankind copes with both global and private disaster is through an art that wrestles with the status of landscape and memory. Descriptions of the natural landscape function as monuments to humankind: '[t]he first stop [on the tour] that was interesting was the Grand Canyon, a natural wonder whose desolate beauty was unaffected by the Sterilization' (Haldeman, 2001, 257). 'For White Hill' landscapes space in terms of its relationship to a human past. That the Grand Canyon is unaffected by the Sterilization points to the autonomous existence of a space that exceeds the human capacity to landscape, and yet the context in which the artists experience this space introduces an aesthetic of loss and desolation. The natural beauty of the Grand Canyon and its independence from the affairs of sentient beings demonstrates aspects of the Asymmetry Thesis. Description of Chicago mentions its '[a]reas of stunning imaginative brilliance next to jury-rigged junk. And everywhere bones, the skeletons of ten million people, lying where they fell' (257). The story's use of the dying Earth narrative alongside the terraforming motif generates nostalgia for a natural and cultural past that has been severed through conflict from the present of the story.

Robert Reed's 'A Place With Shade' (2001 [1995]) and 'A History of

Terraforming' (2010) exemplify this connection between terraforming and reflection on the history of adaptations to Earth. The narrator of 'A Place With Shade' frames terraforming as an extension of the desire to transform the immediate environment, compelling the historical development of increasingly complex technologies. One passage connects humankind's history of altering and remaking physical space, of building and dwelling on Earth, to the sophisticated process of industrial and biological terraforming:

> Terraforming is an ancient profession.
>
> Making your world more habitable began on the Earth itself, with the first dancing fire that warmed its builder's cave; and everything since – every green world and asteroid and comet – is an enlargement on that first cozy cave. A hotter fusion fire brings heat and light, and benign organisms roam inside standardized biomes. For two hundred and ten centuries humans have expanded the Realm, mastering the tricks to bring life to a nearly dead universe. The frontier is an expanding sphere more than twenty light-years in radius – a great peaceful firestorm of life – and to date only one other living world has been discovered. Pitcairn. Alien and violent, and gorgeous. And the basic inspiration for the recent New Traditionalist movement. Pitcairn showed us how bland and domesticated our homes had become, riddled with cliches, every world essentially like every other world. Sad, sad, sad. (Reed, 2001, 199)

Terraforming is seen as an extension of humankind's attempt to make of Earth a suitable human habitation. The bringing of life to a dead universe ultimately results in a superficial existence in which people's lives have been progressively shorn of physical obstacles and challenges to overcome. These challenges are directly connected to a sublime aesthetic principle informed by experience with an alien nature's otherness.

In this story, Locum's pupil Ula sabotages the designed world that he has been contracted to create on her father's (Provo Lei's) planet. Lei disapproves of Locum's profession, considering it 'pretentious and wasteful, this business of building cruel places' (Reed, 2001, 193). Ula's earlier experiments caused the ecological collapse of the planet's original Beringan ecology, which at the time of Locum's visit resembled 'an inspired apartment complex, lovely in every superficial way' (Reed, 2001, 195). The organisms that Ula introduces to this sanitised and calibrated world are a reminder of the otherness of nature and the value judgements that result in the exclusion of specific organisms from this world. In an extension of this environmental ethic to human concerns, Lei tells Locum of Ula's history of neglect and abuse before he adopted her. 'A Place With Shade' suggests that human injustice and the exclusion of elements of nature stem from a disavowal of history. The question of an environmental ethic directed towards microbiological life is indebted to the influence of the 1960s environmental movement and to sf's longstanding engagement with the idea of nature, in both its cosmological and its biological senses.

Locum subscribes to the New Traditionalist (NT) movement, which seeks 'to regain the honesty of the original Earth. Hard winters. Droughts. Violent predation. Vibrant chaos' (Reed, 2001, 197). Another NT principle is that '[t]he fit survive. We build worlds with too much diversity, knowing that some of our creations are temporary. And unworthy' (200). One NT motto is the classic statement '[r]ed of tooth and claw,' which alludes to the Hobbesian view of evolution as violent competition (205). What makes the original Earth special is the 'three billion years of natural selection, amoral and frequently shortsighted ... and wondrous in its beauty, power, and scope ... and how we in the Realm had perfected a stupefying version of that wonder, a million worlds guaranteed to be safe and comfortable for the trillions of souls clinging to them' (205). Locum dislikes the trend of terraforming worlds for ease of living as much as Lei dislikes Locum's NT ethic. Ula, however, is disappointed with Locum's beliefs and goes further, claiming that 'Nature is so much more cruel and honest than you'd ever be' (206).

Ula tells Locum of an incident in which she re-enacted the ecological collapse through freezing of the original deep sea vent communities on Lei's world, where she froze herself along with one such community that she had re-established. Locum can only explain this behaviour as a way for her to 'know how if felt' (Reed, 2001, 209) for a species to undergo ecological collapse. Ula asks Locum, '[h]ow do you think it would feel? Your world is thrown free of your sun, growing cold and freezing over ... nothing you can do about it ... and how would you feel...?' (207), but Locum is unsympathetic because only bacteria perished, '[n]othing sentient' (208). Her recollection of this affective response towards non-sentient life shows a respect for all life that is unusual in earlier terraforming narratives, which tend to concern themselves with the preservation of sentient alien life. She relates a childhood story involving a visit to a new mine where evidence of the original ecology is preserved as 'striations ... how layers of bacteria had grown, by the trillion ... outnumbering the human race,' causing her to cry '[b]ecause they had died' (208). This evocative response is laden with guilt over humankind's assumption of superiority over 'lesser' organisms and represents a significant extension of moral awareness to organisms usually excluded from such ethical concern.

The notion of landscaping as both an intellectual and a physical process is explicitly related to the concept of the scientific model and simulation:

> Terraformers build their worlds at least twice.
>
> The first time it is a model, a series of assumptions and hard numbers inside the best computers; and the second time it is wood and flesh, false sunlight and honest sound. And that second incarnation is never the same as the model. It's an eternal lesson learned by every terraformer, and by every other person working with complexity.
>
> Models fail.
>
> Reality conspires. (Reed, 2001, 204)

The gap between the ideal and the manifest substrates the status of terraformed worlds as models. Within this creative space multiple ideals converge and struggle with one another. As expectation cedes to experience, the complexity of nature resists control and alteration. The terraformed world is related to the inner world of the human mind: '[m] inds are secretive. And subtle. And molding them is never so easy and clear as the molding of mere worlds, I think' (Reed, 2001, 204). Locum remembers an incident at the academy on the day of his graduation: an aged teacher asks him '[w]here do these worlds we build actually live?' Her answer is that they reside '[i]n our minds [...] That's the only place they can live for us, because where else can we live?' (207). These passages reflect a divorce between internal and external worlds; terraforming attempts to make manifest a clash of ideals that seek both intellectual and physical resolution.

Like Williamson's *Terraforming Earth*, 'A History of Terraforming' mobilises vast sweeps of time to explore the relationship between changes to society and the continuity of human nature. This story dramatises the slow struggle towards a sense of belonging and habitation and features life-extending technologies that allow the narrator to focus on the influence of a pivotal individual's life over the course of humankind's expansion throughout the galaxy. The elite-driven shaping of history and the relationship between leaders and followers are central themes, while the past and future are essential to the way individuals approach nature and society. When alternative views of the past are dialogised, the past and future begin to reciprocally shape each other. Knowledge of the past, while not sufficient for an environmental and egalitarian view of others, is a necessary condition for developing an appropriately scaled view of human relationships to the universe.

'A History of Terraforming' features six episodes from the life of the last of the first 'atums,' a name borrowed from Egyptian mythology referring to the 'god whose task it was to finish the unfinished worlds' (Reed, 2010, 81). Examples of pantropy occur alongside terraforming, illustrating the close connection between the two motifs, which for reasons of space, this book has not been able to explore in detail. Human bodily modifications in this story include life-extending technologies, the miniaturisation of bodies, the substitution of crystalline structures for organic minds and a communal databank for the storage of memories. Simon remembers how, when he was a child, how the Martian community's hope for a safe Mars drove the desire to terraform, a motive grounded in an internalisation of Keekok Lee's Asymmetry Thesis and which ultimately leads to a view of terraforming as a battle against nature. Two collapses of the Martian ecology confirm the early atums' distrust of organic solutions and their preference for technological solutions that seemingly offer surer control of the environment. Simon, in his later role as chief atum, follows another philosophy that embraces life as offering the most sustainable structure for terraforming. He couples this to an acceptance of local-scale ecological failures as an expression of this life. After a period of galactic warfare, Simon confronts Naomi, a colleague formerly engaged in a genocidal struggle to ensure that events conform to her vision of suitable finished worlds. This episode leads to Simon's execution of Naomi for war crimes, thus affirming his rejection of the mastery of nature and of societies that his conventionally brilliant and successful colleague extols.

Acceptance of the uncertainty that life represents is prefigured by Simon's childhood entanglement in a debate over terraforming Mars. Like Clayborne in Robinson's *Mars* trilogy, Lily works to preserve the original ecology of Mars. Unlike in *Mars*, microbial life is proven to exist on the planet and some of the community's members are involved in a preservationist project to collect samples for the 'Zoo,' a repository of Martian micro-organisms. Simon views this preservation as 'sad,' arguing that 'Life should be busy' (Reed, 2010, 88). At the zenith of his career he enacts a plan to protect extra-galactic planets from terraformation by a group of atums who had abandoned the galactic society early in Simon's career, thus ensuring the preservation of indigenous life in those environments.

Simon's stance towards extra-galactic terraforming is built on the idea that people should look towards the quality of their lives within their present environments: '[w]e will embark as soon as we can trust our nature and our institutions not to use this migration as an excuse for easy growth and return voyages of conquest [...]. When we have become adults, finally mature and responsible on all occasions' (Reed, 2010, 106). This story declaims against power, celebrates life and asks that humankind mature without losing a sense of wonder towards the universe; in an echo of Laurence Manning's 'The Living Galaxy,' Simon asserts that 'we will not leave this little realm of ours until we are children again. Wide-eyed, enthralled children who know what they have in their hands and hold it with all the care they possess' (106). Simon stands as witness to the struggle that accompanies humankind's colonisation of the galaxy. In his climactic speech, he attempts to leave a legacy of wonder and restraint for the civilisations of the future.

Linda Nagata's 'Goddesses' (2011 [2000]) exemplifies the increased interest in geoengineering themes in the terraforming narratives of this period. 'Goddesses' is told from the perspectives of three people involved in the remediation projects of two corporations, Global Shear in Four Villages, India, and Green Stomp in America, allowing the narrative to map themes across geopolitical boundaries. The narrative questions and attempts to break down assumptions, a principle that Cody of Green Stomp connects to environmental rejuvenation in a manner that echoes Ursula K. Le Guin's use of the wall motif in *The Dispossessed*: '[k]icking apart toxic "non-biodegradable" molecules was a physical thrill. In her mind, it was the same as kicking down the mental walls that fenced people in' (Nagata, 2011, 244–45). One of the walls that the narrative attempts to address 'said [that] technology must eventually lead to apocalypse, whether through war, engineered disease, overpopulation or pollution' (245). 'Goddesses' is holist in its consideration of the role that corporations should play in addressing issues of health, conflict and responsibility, while it also weighs the impact of traditional socio-cultural views directing public concern. Global Shear and Green Stomp are involved in bioremediation, which attempts to tackle the problem of environmental degradation through the socio-economic restructuring of society alongside the physical management of the environment via ecopoietic means. Cody is motivated by a personal history of chemical poisoning that seems to have left her barren, a legacy of corporations with irresponsible policies for chemical disposal. Michael of Global Shear thinks that 'Four Villages was a microcosm of the world and it faced formidable problems – poverty, overpopulation, illiteracy, environmental degradation, and perhaps worst of all, the poison of old ideas – but none of these challenges was insurmountable' (335). 'Goddesses' clearly marks its structural similarity to the modelling aspect of the terraformed world and relates physical poisons to 'the poison of old ideas.'

Debate in 'Goddesses' is spatialised; its structure maps ideas across several domains and contexts. Audio-visual technologies work to foster collaborations between individuals in various continents, thus widening the field of experience available to those with access to these networks. Fundamentalism is critiqued for its inhibition of appropriate adaptations to new social environments, changes that can be traced back to alterations in society's physical environment. Gharia in India and Cody in America are juxtaposed as examples of two types of fundamentalism: '[b]oth of them had let antique expectations twist the balance of their lives' (Nagata, 2011, 323). Cody's inability to adapt her expectation of a future with daughters motivates her divorce from Michael, while the decreasing population of females in Four Villages prompts Gharia to blackmail Michael with the intention of possessing through marriage the final viewpoint character of the narrative, the abused widow Rajban. Michael's superior questions his aid to Rajban, arguing that harbouring a widow breaches traditional etiquette and sows distrust amongst the population. Michael, however, realises that it is with the marginalised women of Four Villages that he should be building trust. In response to his manager's reminder that '[y]ou are there to grow an economy, not to rescue damsels in distress,' Michael argues that '[d]amsels are part of the economy, Karen. Everyone matters and you know it. The more inclusive the system is, the more we all benefit' (335). This economic perspective bridges physical and social parameters to address the limits of corporate social responsibility.

'Goddesses' is a story about ideas, but also about the power of the idea, which is likened to a form of magic. Rajban's personal history is testimony to the mistreatment of women in society, but her narrative trajectory illustrates how the seed of an idea can take root and grow, promising to open spaces where this mistreatment can be contested. Michael's own experience mirrors Rajban's change of consciousness; their interaction begins a process of mutual education and adaptation to new social contexts. Rajban's exposure to the ideas of an all-female group who support each other in business ventures funded by the Southern Banking Alliance culminates first in capitulation to the demands of her husband's family, and then in rebellion against the abuse and the expectations that she is demanded to meet. An ecological image is emblematic of the growth of Rajban's rebellion and establishes a new position for her with regard to her moral environment: '[h]er fists clenched as the seed sprouted in a burst of growth, rooting deep down in her gut and flowering in her brain, thriving on the magic soil of new ideas' (Nagata, 2011, 321). As a gardener able to conjure '[l]ife from lifelessness,' an ability that Rajban thinks of as 'magic ha[ving] flowed into the soil,' she connects this emblem for the power of ideas to take root, grow and change social environments to the physical process of tending a garden (262). This magic is given sf credibility when it is discovered that her only possession, a pouch of dirt, is 'a natural bioremediation culture, a community of microorganisms fine-tuned for the pollutants particular to the soil around Four Villages,' a discovery rooted in history and indigenous knowledge that offers Rajban economic independence (333). Terraforming symbolises this interdependence between the idea and its power to change the world.

Emphasis on the past, salient during this period, reflects on a history of climate change, planetary adaptation and the building of societies and civilisations, but also meta-textually on the history of terraforming in sf, a notion that Reed's 'A History of Terraforming' makes especially clear in its re-use of motifs and ideas from other terraforming stories. These stories reflect elegiacally on colonisation, but dwell on the theme in a more troubled way: the legacy left by other terraforming stories is one that questions and struggles ambivalently with the idea of planetary and societal adaptation. They are less sure of the outcome of terraforming and anxious about the relationship between the treatment of nature and society. As 'Goddesses' shows, a history of environmental degradation from an era of industrial contamination is now firmly integrated into the terraforming megatext. However far sf narratives of terraforming might range, these stories show in microcosm that the tradition comes full circle to reflect back on contemporary Earthbound concerns.

James Cameron's *Avatar* (2009) introduces to a popular audience themes and images central to the terraforming tradition. It revitalises tropes developed in the sf discourse of the 1960s–1970s and frames these countercultural voices in the context of a society that perceives environmental threat in terms of risk, global warming and climate change. Sf discourse acts as a cultural repository; *Avatar*'s narrative is an excavation and re-presentation of tropes developed within the written sf tradition. *Avatar* can not only help to illuminate aspects of terraforming and the Gaia hypothesis, but may prime new audiences to the relevance of earlier narratives to contemporary ecological concerns. Using *Avatar* as a thematic reference for considering the contribution of earlier sf in shaping this tradition makes salient the dialogical aspect of the sf megatext and illuminates the contours of the wider sf tradition from which various themes are appropriated and cinematically re-deployed and thus transformed in new contexts.

Avatar dramatises the conflict between an interplanetary mining corporation and the indigenous alien Na'vi living on the moon Pandora. Set in the year 2154, the narrative follows Sully's experiences as a paraplegic mercenary who accepts an invitation from the Company to take his deceased twin brother's place on the Avatar project. Avatars are expensive constructs modelled on Na'vi physiology; their neurological structure is mapped against a human controller's, thus allowing the remote control of these bodies. By inhabiting the body of his Avatar, Sully is able to move freely on the surface of Pandora, immune to atmospheric toxins and freed from the physical limitations he experiences in his human body. The scientific team employs Avatars to aid in the exploration and research of Pandora and its ecology and to interact with the Na'vi. Peaceful relations are complicated when Colonel Quaritch offers Sully an expensive regenerative treatment for his paraplegia. In return, Quaritch demands military intelligence on the Na'vi for the Company's efforts to displace them and gain access to deposits of the coveted mineral 'unobtainium' that lie beneath the Na'vi settlement Hometree.

James Cameron has acknowledged the cinematic influence on *Avatar* of such films as *At Play in the Fields of the Lord*, *Dances With Wolves* and *The Emerald Forest*, explaining that 'I just gathered all this stuff in and then [...] looked at it through the lens of science fiction' (see Boucher, 2009). Cameron also acknowledges a debt to sf literature when he notes that '[the idea for *Avatar*] came from all the science-fiction books I read when I was a kid and it just gestated over time' (see Hiscock, 2009). In the tradition of H. Rider Haggard's *Allan Quatermain* (1885–1927) and Edgar Rice Burroughs's *Barsoom* series (1912–1964), Cameron calls *Avatar* 'an old-fashioned jungle adventure with an environmental conscience. It aspires to a mythic level of storytelling' (see Rampton, 2009). Other commentators have noted strong similarities to other works of sf literature, including Poul Anderson's short story 'Call Me Joe' (1989 [1957]), in which a paraplegic scientist employs technology operating on the basis of telepathy to inhabit the body of a five-foot slate-blue centaur-like creature designed for the purpose of colonising Jupiter (Westfahl, 2009). Anderson's novel *The Avatar* (1978) is also of peripheral interest for its title and use of the avatar motif, its vignettes focalised through animal and plant perspectives and its exploration of nature at ecological and cosmological scales.

Cameron explains that in *Avatar* '[w]e're telling the story of what happens when a technologically superior culture comes into a place with a technologically inferior indigenous culture and there are resources there that they want.' This colonial narrative is set against what Cameron explains is 'a love story about an awakening of perception through the other person' (see Hiscock, 2009). This statement highlights the two main narrative threads of *Avatar*. The first is the colonial acquisition of new territory seen only as a repository of resources, which is concomitant with the displacement of indigenous cultures. The second trajectory involves a dramatisation of Sully's change of perspective towards the external world. This change is directly initiated by an affective response towards an environment conceived of as 'whole' and 'connected,' elements of a belief system based upon the Na'vi's spiritual understanding of an entity they call Eywa. *Avatar* combines a critique of imperial colonisation and exploitation with the portrayal of an alternative complex of religio-spiritual and affective perspectives and attitudes towards nature.

Avatar opens with a dream of flying, which is eventually fulfilled when Sully passes his initiation test by taming a dragon-like Ikran and later a Toruk, a legendary feat accomplished by only five Na'vi. According to Na'vi myth, this act legitimises his position as leader, granting him the respect necessary to unite them and lead them to war against the colonists. Much criticism has been levelled against this theme because of its perpetuation of the assumption that victims of colonial oppression can only be delivered from such exploitation by one of the colonisers. Vera and Gordon suggest that '[t]his movie is supposedly set on the distant planet Pandora, but it really takes place close to home, for *it opens up the Pandora's box of the American racial unconscious*.' They argue that in *Avatar* '[t]he racial masquerade is another fantasy solution to white guilt in which the white hero crosses over and pretends to be black or native American' (Vera and Gordon, 2009). This reading addresses guilt as a response to the historical exploitation of cultures and nature. Like the dreamer gods of Le Guin's *The Word for World Is Forest*, *Avatar* presents the spectator with tendencies that are implicit in American culture. *Avatar*'s generic influences enter into dialogue with the film to offer the viewer a vantage from which to question the politics of its use of religio-spiritual themes and the relationship to nature they express.

Na'vi spiritualism centres around a being they call Eywa. Dr Norm Spellman explains to Jake, '[w]ho's Eywa? Only their deity. Their goddess made up of all living things. Everything they know.' Na'vi belief in the Great Mother figure Eywa draws from the quasi-pantheistic popularisations of Lovelock's Gaia hypothesis, an intersection between Gaia and terraforming that has one of its roots in the environmental politics of the 1960s. Grace Augustine, the leading colonial scientist, draws from the scientific basis of Lovelock's hypothesis to explain the

Na'vi's connection to their forest environment. This blend of mysticism and science is characteristic of treatments of the living world motif and of Lovelock's early discussion of Gaia. The appearance of floating seeds from the Tree of Souls, what the Na'vi Neytiri explains are '[s]eeds of the sacred tree. Very pure spirits,' stops her from killing Sully at first sight and compels her to bring him to Hometree, the village of the Omaticayan clan of Na'vi. Neytiri interprets the appearance of these seeds as a sign from Eywa. Later, she tells Sully, '[o]ur Great Mother does not take sides, Jake. She protects only the balance of life.' Images of a Great Mother protecting the 'balance of life' draw from a pseudo-pantheistic vision of energy continuously flowing through discrete bodies of organic life. During the sequence in which Neytiri teaches Sully to behave and think as a Na'vi, he records in his video diary that Neytiri is 'always going on about the flow of energy, the spirits of animals,' and that 'I'm trying to understand this deep connection the people have to the forest. She talks about a network of energy that flows through all living things. She says all energy is only borrowed, and one day you have to give it back'.

The polarisation of the Na'vi and the Company in *Avatar* is a variant of a mythic archetype, one that sets the Na'vi's ecological spirituality against the Company's narrow, 'rationalist' instrumental view of nature. The colonists view the Na'vi as technologically primitive and irrational and interpret these characteristics as the source of their resistance to the mining operation. Grace contests this view and presents the scientific basis for the Na'vi worship of Eywa:

> What we think we know is that there's some kind of electro-chemical communication between the roots of the trees, like the synapses between neurons. And each tree has ten to the fourth connections to the trees around it. And there are ten to the twelfth trees on Pandora.
>
> [...]
>
> It's more connections than the human brain. Get it? It's a network. It's a global network and the Na'vi can access it, they can upload and download data. Memories at sites like the one you just destroyed.

The site Grace refers to is the 'Tree of Voices,' under which the Na'vi dead are buried and their memories absorbed into the wider Gaian system. This description is joined to the language of computing and networks, which separates it from the organic-spiritual epistemology by which this trope was rendered in the 1960s–1970s. The bond through which the Na'vi trade information resembles a braid falling from the back of their heads; through this braid they are able to connect to other life forms. This image echoes that of the plugs in *The Matrix* (1999) and blurs the boundary between ideas of nature and technology. Such crossings between organic and technological discourse are already implicit in earlier Gaian images. *Avatar*'s portrayal of nature spirituality draws from the language of sf literature and operates as a compilation of ecologic sf traditions that have gained wider currency in a culture that, since the 1980s, has been developing ways of relating to nature in response to the threat of global climate change.

Coda

Terraforming has continued to infiltrate scientific understanding of environmental change on Earth since 2000. The real-world example of geoengineering embodied by Ascension Island in the Atlantic, where Charles Darwin and Joseph Hooker began a project in 1850 to transform the arid volcanic island into its present-day forested environment, has been acknowledged as an important example of directed environmental change and a premonition of terraforming (Falcon-Lang, 2010). In 2012, the three-part BBC documentary *How to Grow a Planet* drew on terraforming and Gaian themes to frame its discussion of the evolution of Earth's planetary environment. This book has raised the relationship between science and sf on many occasions, but the motif of terraforming offers the potential for a detailed analysis of the feedback between these two areas of discourse and of the language used to communicate science to the wider public.

Along with the anthology *Worldmakers: Sf Adventures in Terraforming* (2001), Gardner Dozois has edited a companion volume dealing with post and transhumanism: *Supermen: Tales of the Posthuman Future* (2002). This book has acknowledged the relationship between terraforming and pantropy and has to some extent considered its importance, but a closer examination of the significance of bodily adaptation and planetary colonisation, along with themes relevant to human–animal studies, would suitably address an area necessarily given less emphasis in these pages. Analysis of works such as Joan Slonczewski's *Elysium* sequence (1986; 1993; 2009 [1998]) and John Brunner's *Bedlam Planet* (1982 [1968]) and *The Dramaturges of Yan* (1974 [1972]) would be central to such a study. Likewise, drawing links between terraforming and related sf themes that explore ecological ideas would also prove profitable; relevant narratives might include ecotastrophe, natural disaster and dying world stories, all of which are united by the discourse of apocalypse.

Terraforming has also increasingly infiltrated film and television, featuring as the renegade astronauts' goal in the two-part 'Space Race' episode of Adam Reed's *Archer* (2012a; 2012b), as a rumoured corporate agenda in Ridley Scott's *Prometheus* (2012) and as a weapon of mass destruction in Zack Snyder's *Man of Steel* (2013). Although this book has considered William Cameron Menzies's *Things to Come* (1936) and Cameron's *Avatar* (2009), further research into the terraforming motif in film and television would be illuminating. Examples of relevant texts include David Lynch's *Dune* (1984), the *Star Trek: The Next Generation* episodes 'Home Soil' (1988) and 'Family' (1990), the movie *Star Trek II: The Wrath of Khan* (1982), Joss Whedon's *Firefly* (2002–2003) series and the *Firefly* film *Serenity* (2005), and anime such as Keiichi Sugiyama's *Origin: Spirits of the Past* (*Gin-iro no kami no Agito*, 2006). This attention to terraforming in other media can be extended to computer gaming: Lovelock consulted on the design of the simulation game *SimEarth: The Living Planet* (1990). Considering also such real-time strategy games as *Dune II: The Building of a Dynasty* (1992) and *Sid Meier's Alpha Centauri* (1999), and such 3D shooters as *Red Faction* (2001), there is ample potential for an investigation of terraforming in multiple media.

This book has necessarily surveyed and examined only the dominant trends in the development of the terraforming motif. In addition to the many novels, short stories and poems that this book has been unable to examine, further study might consider the feedback between science and sf and attempt a more detailed analysis of the motif's appearance in other textual formats, including film, television, art and games, undertakings that will undoubtedly uncover subtler or more localised trends that may yield new insight into the development of the terraforming narrative and its place as a master motif for contemporary environmental thought. Terraforming and geoengineering have infiltrated wider popular culture and are appearing with increasing regularity in the discourse of news media and the science of climate change and mitigation. Sf has greatly influenced the way in which climate change is imagined today, providing one important reason for examining the theme in sf literature. This attention to the mode allows us to explore the assumptions that underlie the way we conceive of our relationship to the spaces we inhabit, and our view of the future we wish to shape out of the cultural traditions and raw materials that surround us. As debate about geoengineering as a form of climate change mitigation grows increasingly important and some real-world experiments are attempted, it pays to explore the thought that has shaped the way we view terraforming and geoengineering and their place in our cultural landscape. Our extrapolations of the present would gain greater power if that present were informed by the literary experiments that probe the values that we attach to the motif of terraforming within our cultures. Literature and film offers a repository in itself, a library of investigations into the intersections between science and society. In the context of our contemporary climate crisis, this investigation is more urgent than ever.

Works Cited

Primary Works Cited

Aldiss, Brian and Roger Penrose, *White Mars; or, The Mind Set Free, a 21st Century Utopia* (London: Little, Brown, 1999).

Allaby, Michael and James Lovelock, *The Greening of Mars* (New York: St. Martin's Press, 1984).

Allen, Cory (dir.), 'Home Soil,' *Star Trek: The Next Generation* (Paramount: 44 mins, 1988).

Anderson, Poul, *The Avatar* (New York: Berkley Publishing Corporation, 1978).

—, 'The Big Rain,' in Gardner Dozois (ed.), *Worldmakers: Sf Adventures in Terraforming* (New York: St. Martin's Griffin, 2001 [1954]), pp. 1–49.

—, 'Call Me Joe,' in Martin Greenberg (ed.), *Isaac Asimov Presents the Great Sf Stories: 19* (New York: Daw Books, 1989 [1957]), pp. 103–48.

—, 'Sister Planet,' in Leo Margulies (ed.), *Get Out of My Sky: Three Short Novels of Science Fiction* (New York: Crest, 1960 [1959]), pp. 87–128.

—, *The Snows of Ganymede* (New York: Ace, 1958 [1955]).

Asimov, Isaac, *Foundation* (London: Panther, 1985 [1951]).

—, 'The Martian Way,' in Ben Bova (ed.), *The Science Fiction Hall of Fame*, vol. 2B (New York: Tom Doherty Associates, 1974 [1952]), pp. 1–45.

Babenco, Hector (dir.), *At Play in the Fields of the Lord* (Universal Pictures: 189 mins, 1992).

Bacon, Francis, *The New Atlantis* (Adelaide: University of Adelaide Library, 2008 [1627]).

Bellamy, Edward, *Looking Backward, 2000–1887* (Oxford: Oxford University Press, 2007 [1888]).

Bester, Alfred, 'Adam and No Eve,' *Astounding Science Fiction*, 28.1 (1941), 35–45.

Blake, William, *The Marriage of Heaven and Hell* (Oxford: Oxford University Press, 1975 [1790]).

Blish, James, *The Seedling Stars* (London: Gollancz, 2001 [1956]).

Boorman, John (dir.), *The Emerald Forest* (Christel Films: 114 mins, 1985).

Bradbury, Ray, 'Here There Be Tygers,' in *R Is for Rocket* (London: Panther, 1972 [1951]), pp. 119–33.

—, *The Martian Chronicles* (New York: Doubleday, 1958 [1950]).

Brin, David, *Earth* (London: Futura, 1990).
Brunner, John, *Bedlam Planet* (New York: Ballantine Books, 1982 [1968]).
—, *The Dramaturges of Yan* (London: New English Library, 1974 [1972]).
—, *The Sheep Look Up* (New York: Harper and Row, 1972).
—, *Stand on Zanzibar* (London: Millennium, 1999 [1968]).
Burroughs, Edgar Rice, *A Princess of Mars* (New York: Ballantine, 1980 [1912]).
Butler, Samuel, *Erewhon; or, Over the Range* (London: Penguin, 1977 [1872]).
Callenbach, Ernest, *Ecotopia: A Novel About Ecology, People and Politics in 1999* (London: Pluto Press, 1978 [1975]).
Cameron, James (dir.), *Avatar* (20th Century Fox: 162 mins, 2009).
Clarke, Arthur C., 'Before Eden,' in Gardner Dozois (ed.), *Worldmakers: Sf Adventures in Terraforming* (New York: St. Martin's Griffin, 2001 [1961]), pp. 60–68.
—, *The Sands of Mars* (London: Sidgwick & Jackson, 1976 [1951]).
Costner, Kevin (dir.), *Dances With Wolves* (Tig Productions: 181 mins, 1990).
Doyle, Arthur Conan, 'When the World Screamed,' *Forgotten Futures III* (2000 [1928]) <http://www.forgottenfutures.com/game/ff3/wscream.htm> (accessed March 2011).
Dozois, Gardner (ed.), *Supermen: Tales of the Posthuman Future* (New York: St. Martin's Griffin, 2002).
— (ed.), *Worldmakers: Sf Adventures in Terraforming* (New York: St. Martin's Griffin, 2001).
Fearn, John Russell, 'Earth's Mausoleum,' *Astounding Stories*, 15.3 (1935), 52–87.
Gerrold, David, *Moonstar Odyssey* (New York: New American Library, 1977).
Haggard, H. Rider, *King Solomon's Mines* (Oxford: Oxford University Press, 1989 [1885]).
Haldeman, Joe, 'For White Hill,' in Gardner Dozois (ed.), *Worldmakers: Sf Adventures in Terraforming* (New York: St. Martin's Griffin, 2001 [1995]), pp. 245–76.
Hamilton, Edmond, 'The Earth-Brain,' in *The Horror on the Asteroid and Other Tales of Planetary Horror* (London: Philip Allan, 1936 [1932]), pp. 129–83.
Heinlein, Robert, *Farmer in the Sky* (London: Pan Books, 1967 [1950]).
—, *The Moon Is a Harsh Mistress* (London: Gollancz, 2001 [1966]).
—, *Stranger in a Strange Land* (New York: Putnam, 1961).
Herbert, Frank, *Children of Dune* (Kent: New English Library, 1976).
—, *Dune* (Kent: New English Library, 1965).
—, *Dune Messiah* (Kent: New English Library, 1971 [1969]).
Hudson, W.H., *A Crystal Age* (London: Duckworth, 1927 [1887]).
Jefferies, Richard, *After London; or, Wild England* (2004 [1885]) <http://www.gutenberg.org/ebooks/13944> (accessed September 2014).
Kipling, Rudyard, 'With the Night Mail: A Story of 2000 AD,' *Forgotten Futures* (2009 [1905]) <http://www.forgottenfutures.com/game/ff1/night.htm> (accessed December 2011).
Landau, Les (dir.), 'Family,' *Star Trek: The Next Generation* (Paramount: 60 mins, 1990).

Landis, Geoffrey A., 'Ecopoesis,' in Gardner Dozois (ed.), *Worldmakers: Sf Adventures in Terraforming* (New York: St. Martin's Griffin, 2001 [1994]), pp. 311–41.

Lang, Fritz (dir.), *Metropolis* (Eureka Video: 153 mins, 1997 [1927]).

Le Guin, Ursula K., *Always Coming Home* (Toronto: Bantam, 1985).

—, *The Dispossessed* (London: Millennium, 1999 [1974]).

—, 'Vaster Than Empires and More Slow,' in *The Wind's Twelve Quarters* (London: Granada, 1982 [1971]), pp. 25–59.

—, *The Word for World Is Forest* (New York: Berkley Publishing, 1976 [1972]).

Leinster, Murray, 'First Contact,' *Crossroads: The Library* (2004 [1945]) <http://thevirtualcommunity.yuku.com/topic/1185/First-Contact-Murray-Leinster#.VGlC7PmsVyU> (accessed October 2007).

—, 'The Lonely Planet,' *Thrilling Wonder Stories*, 35.2 (1949), 80–97.

Lem, Stanislaw, *Solaris* (London: Faber & Faber, 2003 [1961]).

Lovecraft, H.P., 'At the Mountains of Madness,' *The H.P. Lovecraft Archive* (2009a [1936]) <http://www.hplovecraft.com/writings/texts/fiction/mm.asp> (accessed January 2012).

—, 'The Colour Out of Space,' *The H.P. Lovecraft Archive* (2009c [1927]) <http://www.hplovecraft.com/writings/texts/fiction/cs.asp> (accessed January 2010).

—, 'The Dreams in the Witch House,' *The H.P. Lovecraft Archive* (2009d [1933]) <http://www.hplovecraft.com/writings/fiction/dwh.asp> (accessed January 2011).

Lynch, David (dir.), *Dune* (De Laurentiis: 137 mins, 1984).

McKenna, Richard, 'Hunter, Come Home,' in Gardner Dozois (ed.), *Worldmakers: Sf Adventures in Terraforming* (New York: St. Martin's Griffin, 2001 [1963]), pp. 69–98.

—, 'The Night of Hoggy Darn,' in *Tomorrow ×4* (Connecticut: Gold Medal Books, 1964 [1958]), pp. 9–58.

Manning, Laurence, 'The Living Galaxy,' *Wonder Stories*, 6.4 (1934), 436–44.

—, *The Man Who Awoke* (New York: Ballantine, 1979 [1933]).

Marvell, Andrew, 'To His Coy Mistress,' *Luminarium: Anthology of English Literature* (2009 [1681]) <http://www.luminarium.org/sevenlit/marvell/coy.htm> (accessed February 2012).

Melville, Herman, *Moby-Dick; or, The Whale* (Oxford: Oxford University Press, 1963 [1851]).

Menzies, William Cameron (dir.), *Things to Come* (London Films: 117 mins, 1936).

Meyer, Nicholas (dir.), *Star Trek II: The Wrath of Khan* (Paramount: 113 mins, 1982).

Miller, Walter M., 'Crucifixus Etiam,' in *The View From the Stars* (Hertfordshire: Panther, 1973 [1953]), pp. 58–78.

Morris, William, *News From Nowhere; or, An Epoch of Rest: Being Some Chapters From a Utopian Romance* (Oxford: Oxford University Press, 2003 [1890]).

Nagata, Linda, 'Goddesses,' in *Goddesses & Other Stories* (Kula: Mythic Island Press, 2011 [2000]), pp. 233–336.

Pohl, Frederik, *Man Plus* (London: Gollancz, 2000 [1976]).

—, *Mining the Oort* (New York: Ballantine Books, 1992).
Pohl, Frederik and C.M. Kornbluth, *The Space Merchants* (New York: Random House, 1974 [1953]).
Reed, Adam (dir.), 'Space Race: Part I,' *Archer* (FX / Floyd County / Radical Axis: 22 mins, 2012a).
—, 'Space Race: Part II,' *Archer* (FX / Floyd County / Radical Axis: 22 mins, 2012b).
Reed, Robert, 'A History of Terraforming,' *Asimov's Science Fiction Magazine*, 34.7 (2010), 74–106.
—, 'A Place With Shade,' in Gardner Dozois (ed.), *Worldmakers: Sf Adventures in Terraforming* (New York: St. Martin's Griffin, 2001 [1995]), pp. 193–220.
Robinson, Kim Stanley, *2312* (London: Orbit, 2012).
—, *Blue Mars* (London: Voyager, 1996a).
—, *Fifty Degrees Below* (London: Harper Collins, 2005).
—, *Forty Signs of Rain* (London: Harper Collins, 2004).
—, *Future Primitive: The New Ecotopias* (New York: Tor, 1997 [1994]).
—, *The Gold Coast* (London: Futura, 1989 [1988]).
—, *Green Mars* (London: Voyager, 1996b [1993]).
—, *The Martians* (London: Voyager, 2000 [1999]).
—, *Pacific Edge* (London: Unwin Hyman, 1990).
—, *Red Mars* (London: Voyager, 1996c [1992]).
—, *Sixty Days and Counting* (London: Harper Collins, 2007).
Sargent, Pamela, *Child of Venus* (New York: EOS, 2001a).
—, 'Dream of Venus,' in Gardner Dozois (ed.), *Worldmakers: Sf Adventures in Terraforming* (New York: St. Martin's Griffin, 2001b [2000]), pp. 394–416.
—, *Venus of Dreams* (London: Bantam, 1989a [1986]).
—, *Venus of Shadows* (New York: Doubleday, 1989b [1988]).
Scott, Ridley (dir.), *Prometheus* (Scott Free / Brandywine: 124 mins, 2012).
Shelley, Mary, *Frankenstein; or, The Modern Prometheus* (Oxford: Oxford University Press, 1998 [1818]).
—, *The Last Man* (Peterborough, Ontario: Broadview Press, 1996 [1826]).
Shiel, M.P., *The Purple Cloud, Project Gutenberg* (2004 [1901]) <http://www.gutenberg.org/files/11229/11229-8.txt> (accessed December 2009).
Shoolingin-Jordan, Nick and Nigel Walk (dir.), *How to Grow a Planet* (BBC: 60 mins, 2012).
Slonczewski, Joan, *The Children Star* (Rockville, MD: Phoenix Pick, 2009 [1998]).
—, *Daughter of Elysium* (New York: William Morrow and Co., 1993).
—, *A Door Into Ocean* (New York: Tom Doherty Associates, 1986).
Snyder, Zack (dir.), *Man of Steel* (Warner Bros: 143 mins, 2013).
Spinrad, Norman, *Greenhouse Summer* (Golden, CO: ReAnimus Press, 2013 [1999]).
Stapledon, Olaf, *Last and First Men* (Harmondsworth: Penguin, 1966 [1930]).
—, *Star Maker* (Middletown, CT: Wesleyan University Press, 2004 [1937]).
Stewart, George R., *Earth Abides* (London: Transworld Publishers, 1961 [1949]).

Sugiyama, Keiichi (dir.), *Origin: Spirits of the Past* (*Gin-iro no kami no Agito*) (Production Committee / Gonzo / Media Factory: 95 mins, 2006).

Thoreau, Henry David, *Walden, Project Gutenberg* (2013 [1854]) <http://www.gutenberg.org/ebooks/205> (accessed September 2015).

Turner, Frederick, *A Double Shadow* (New York: Putnam, 1978).

—, *Genesis: An Epic Poem, Frederick Turner's Blog* (1988) <http://frederickturnerpoet.com/?page_id=166> (accessed November 2014).

—, *The New World: An Epic Poem* (Princeton: Princeton University Press, 1985).

Varley, John, 'Retrograde Summer,' in Gardner Dozois (ed.), *Worldmakers: Sf Adventures in Terraforming* (New York: St. Martin's Griffin, 2001 [1975]), pp. 118–33.

Volition, Inc. (developer), *Red Faction* (Volition, Inc., 2001).

Wachowski, Andy and Lana Wachowski (dirs), *The Matrix* (Groucho II Film Partnership: 136 mins, 1999).

Wells, H.G., *The Shape of Things to Come* (London: Corgi, 1967 [1933]).

—, *The War of the Worlds, Project Gutenberg* (2004 [1898]) <http://www.gutenberg.org/cache/epub/36/pg36.txt> (accessed March 2011).

—, 'When the Sleeper Wakes,' *Project Gutenberg* (2008 [1899]) <http://www.gutenberg.org/files/775/775-h/775-h.htm> (accessed April 2012).

Whedon, Joss (dir.), *Firefly* (Mutant Enemy / 20th Century Fox, 2002–2003).

—, *Serenity* (Universal / Barry Mendel: 119 mins, 2005).

White, James, 'Major Operation,' in *Major Operation* (New York: Ballantine Books, 1971a), pp. 136–83.

—, 'Meatball,' in *Major Operation* (New York: Ballantine Books, 1971b [1969]), pp. 103–35.

Williamson, Jack, 'Born of the Sun,' *Astounding Stories*, 12.1 (1934), 10–38.

—, 'Collision Orbit,' in *Seventy-Five: The Diamond Anniversary of a Science Fiction Pioneer* (Michigan: Haffner Press Oak, 2004 [1942]), pp. 216–77.

—, *Terraforming Earth* (New York: Tom Doherty Associates, 2003 [2001]).

Wright, Will (creator), *SimEarth: The Living Planet* (Maxis, 1990).

Zamyatin, Yevgeny, *We* (London: Penguin, 1993 [1924]).

Secondary Works Cited

Aldiss, Brian and David Wingrove, *Trillion Year Spree: The History of Science Fiction* (London: Stratus, 2001).

Ashley, Mike, *The Time Machines: The Story of the Science-Fiction Pulp Magazines From the Beginning to 1950* (Liverpool: Liverpool University Press, 2000).

Attebery, Brian, 'Science Fiction, Parable, and Parabolas,' *Foundation: The International Review of Science Fiction*, 34.95 (2005), 7–22.

Attebery, Brian and Veronica Hollinger, *Parabolas of Science Fiction* (Middletown, CT: Wesleyan University Press, 2013).

Baer, Steve, *Dome Cookbook* (Corrales, NM: Trial and Error, 1996 [1967]).

Bakhtin, M.M., edited by Michael Holquist, *The Dialogic Imagination: Four Essays*, trans. by Caryl Emerson and Michael Holquist (Austin: University of Texas Press, 2002).

Bardini, Thierry, 'Decompicultures: Decomposition of Culture and Cultures of Decomposition,' *Green Letters,* 18.1 (2013), 9–21.

—, *Douglas Engelbart, Coevolution, and the Origins of Personal Computing* (Stanford: Stanford University Press, 2000).

—, *Junkware* (Minneapolis: University of Minnesota Press, 2011).

Baudrillard, Jean, *The Illusion of the End,* trans. by Chris Turner (Oxford: Blackwell, 1994).

Bennett, Jane, *Vibrant Matter: A Political Ecology of Things* (Durham: Duke University Press, 2010).

Ben-Tov, Sharona, *The Artificial Paradise: Science Fiction and American Reality* (Michigan: University of Michigan Press, 1995).

Berry, Adrian, *The Next Ten Thousand Years: A Vision of Man's Future in the Universe* (London: Hodder and Stoughton/Coronet, 1976).

Boucher, Geoff, 'James Cameron: Yes, "Avatar" is "Dances With Wolves" in Space ... Sorta,' *Los Angeles Times* (2009) <http://latimesblogs.latimes.com/herocomplex/2009/08/james-cameron-the-new-trek-rocks-but-transformers-is-gimcrackery.html> (accessed December 2010).

Bourget, Jason, 'Biological Determinism, Masculine Politics and the Failure of Libertarianism in Robert A. Heinlein's *The Moon Is a Harsh Mistress,' Foundation: The International Review of Science Fiction,* 104 (2008), 10–22.

Bramwell, Anna, *Ecology in the 20th Century: A History* (London: Yale University Press, 1990).

Brand, Stuart, *Space Colonies* (New York: Penguin, 1977).

—, *Whole Earth Catalog* (1968) <http://www.wholeearth.com/issue-electronicedition.php?iss=1010> (accessed April 2012).

Broderick, Damien, *Reading by Starlight: Postmodern Science Fiction* (London: Routledge, 1995).

Buhle, Paul, 'Kim Stanley Robinson: Science Fiction Socialist,' *Monthly Review: An Independent Socialist Magazine,* 54.3/4 (2002), 87–90.

Burke, Edmund, 'A Philosophical Enquiry Into the Origin of Our Ideas of the Sublime and Beautiful,' in David Womersley (ed.), *A Philosophical Enquiry Into the Sublime and Beautiful and Other Pre-Revolutionary Writings* (London: Penguin, 1998), pp. 49–119.

Burling, William J., 'The Theoretical Foundation of Utopian Radical Democracy in Kim Stanley Robinson's *Blue Mars,' Utopian Studies,* 16.1 (2005), 75–96.

Carlson, Allen, 'Aesthetic Appreciation, and Knowledge,' *Journal of Aesthetics and Art Criticism,* 53.4 (1995), 393–400.

—, 'Aesthetic Judgement, and Objectivity,' *Journal of Aesthetics and Art Criticism,* 40.1 (1981), 15–27.

—, 'Appreciation and the Natural Environment,' *Journal of Aesthetics and Art Criticism,* 37.3 (1979), 267–75.

Carson, Rachel, *Silent Spring* (Boston: Houghton Mifflin, 2002).

Cooke, Sebastian, 'Sebastian Cooke Talks to Kim Stanley Robinson,' *Eidolon* (1995) <http://eidolon.net/eidolon_magazine/issue_13/13_stan.htm> (accessed November 2014).

Crossley, Robert, *Imagining Mars: A Literary History* (Middletown, CT: Wesleyan University Press, 2011).

Csicsery-Ronay, Jr, Istvan, 'The Book Is the Alien: On Certain and Uncertain Readings of Lem's *Solaris,' Science Fiction Studies,* 12.35 (1985), 6–21.

Daly, Erin Moore and Robert Frodeman, 'Separated at Birth, Signs of Rapprochement: Environmental Ethics and Space Exploration,' *Ethics & the Environment,* 13.1 (2008), 135–51.

de Chardin, Pierre Teilhard, *The Phenomenon of Man,* trans. by Bernard Wall (London: Collins, 1965).

de Luce, Judith, '*Genesis: An Epic Poem*: Inventionist Ecology in Iambic Pentameter,' *The Humanist,* 53 (1993), 19–22.

Delany, Samuel R., *The Jewel Hinged Jaw: Notes on the Language of Science Fiction* (New York: Dragon Press, 1977).

Dumiak, Michael, 'Turning the Red Planet Green,' *Cosmos: Mars Special,* 18 (2007), 62–65.

Dunne, J.W., *An Experiment With Time* (London: Papermac, 1981).

Ehrlich, Paul R., *The Population Bomb* (London: Ballantine, 1971).

'Eldritch, adj.', *Online Oxford English Dictionary* (2015) <http://www.oed.com/view/Entry/60208?redirectedFrom=eldritch#eid> (accessed February 2015).

Empson, William, *Some Versions of Pastoral* (London: Chatto & Windus, 1950).

Engel, Jill, 'Letters From Upstate New York: A Correspondence with Pamela Sargent,' *Nova Express,* 3.3 (1991), 5–9, 19–24.

Falcon-Lang, Howard, 'Charles Darwin's Ecological Experiment on Ascension Island,' *BBC News* (2010) <http://www.bbc.co.uk/news/science-environment-11137903> (accessed November 2012).

Fogg, Martyn J., *Terraforming: Engineering Planetary Environments* (Warrendale: Society of Automotive Engineers, 1995).

Franko, Carol, 'The Density of Utopian Destiny in Robinson's *Red Mars,*' *Extrapolation,* 38.1 (1997), 57–65.

—, 'Working the "In-Between": Kim Stanley Robinson's Utopian Fiction,' *Science Fiction Studies,* 21 (1994), 191–211.

George, Henry, *Progress and Poverty, Library of Economics and Liberty* (New York: Doubleday, 1920 [1879]) <http://www.econlib.org/library/YPDBooks/George/grgPP.html> (accessed May 2012).

Godlovitch, Stan, 'Carlson on Appreciation,' *Journal of Aesthetics and Art Criticism,* 55.1 (1997), 53–55.

—, 'Icebreakers: Environmentalism and Natural Aesthetics,' *Journal of Applied Philosophy,* 11.1 (1994), 15–30.

Gough, Noel, 'Playing With Wor(l)ds: Science Fiction as Environmental Literature,' in Patrick D. Murphy (ed.), *Literature of Nature: An International Sourcebook* (London: Fitzroy Dearborn, 1998), pp. 409–14.

Hailwood, Simon A., *How to Be a Green Liberal: Nature, Value and Liberal Philosophy* (Chesham: Acumen, 2004).

—, 'Landscape, Nature, and Neopragmatism,' *Environmental Ethics,* 29.2 (2007), 131–49.

Haldane, J.B.S., 'The Last Judgement,' in *Possible Worlds and Other Essays* (London: Chatto & Windus, 1927), pp. 287–312.

Harrison, Niall, 'Writing Climate Change: A Round Table Discussion,' *Strange Horizons* (2012) <http://www.strangehorizons.com/2012/20120227/harrison-a.shtml> (accessed November 2014).

Hassler, Donald M., 'Ambivalences in the Venus of Pamela Sargent,' *Extrapolation*, 38.2 (1997), 150–56.

Haynes, Robert, '1: Ecce Ecopoiesis: Playing God on Mars,' in Don MacNiven (ed.), *Moral Expertise: Studies in Practical and Professional Ethics* (London: Routledge, 1990), pp. 161–83.

Heise, Ursula K., 'Reduced Ecologies: Science Fiction and the Meanings of Biological Scarcity,' *European Journal of English Studies*, 16.2 (2012), 99–112.

—, *Sense of Place and Sense of Planet: The Environmental Imagination of the Global* (Oxford: Oxford University Press, 2008).

Herbert, Frank, 'Dune Genesis,' *Dune: The Official Website* (2007) <http://www.frankherbert.org/news/genesis.html> (accessed February 2012).

—, 'They Stopped the Moving Sands,' in *The Road to Dune* (New York: Tor, 2005), pp. 203–10.

Heyd, Thomas, 'Aesthetic Appreciation and the Many Stories About Nature,' *British Journal of Aesthetics*, 41.2 (2001), 125–37.

Hiscock, John, 'James Cameron Interview for *Avatar*,' *Telegraph Online* (2009) <http://www.telegraph.co.uk/culture/film/6720156/James-Cameron-interview-for-Avatar.html> (accessed February 2015).

Hughes, J. Donald, *An Environmental History of the World: Humankind's Changing Role in the Community of Life* (London: Routledge, 2009).

Jameson, Fredric, '"If I Find One Good City I Will Spare the Man": Realism and Utopia in Kim Stanley Robinson's *Mars* Trilogy,' in Patrick Parrinder (ed.), *Learning From Other Worlds: Estrangement, Cognition and the Politics of Science Fiction and Utopia* (Liverpool: Liverpool University Press, 2000), pp. 208–32.

—, 'World Reduction in Le Guin,' in *Archaeologies of the Future: The Desire Called Utopia and Other Science Fictions* (London: Verso, 2005), pp. 267–80.

Johns-Putra, Adeline, *The History of the Epic* (Hampshire: Palgrave Macmillan, 2006).

Kauffman, Stuart A., *Investigations* (Oxford: Oxford University Press, 2000).

Kirk, Andrew G., *Counterculture Green: The Whole Earth Catalog and American Environmentalism* (Lawrence, KS: University Press of Kansas, 2007).

Kolodny, Annette, *The Lay of the Land: Metaphor as Experience and History in American Life and Letters* (Chapel Hill: University of North Carolina Press, 1975).

Kroeber, Theodora, *Ishi in Two Worlds: A Biography of the Last Wild Indian in North America* (Berkeley: University of California Press, 1973 [1961]).

Lease, Gary and Michael E. Soulé, *Reinventing Nature?: Responses to Postmodern Deconstruction* (Washington, DC: Island Press, 1995).

Lee, Keekok, 'Awe and Humility: Intrinsic Value in Nature – Beyond an Earthbound Environmental Ethics,' *Royal Institute of Philosophy Supplement*, 36 (1994), 89–101.

Lem, Stanislaw, 'On Stapledon's *Star Maker*,' *Science Fiction Studies*, 14.41 (1987), 1–8.

Leopold, Aldo, 'The Land Ethic,' in D. Schmidtz and E. Willott (eds), *Environmental Ethics: What Really Matters, What Really Works* (Oxford: Oxford University Press, 2002), pp. 27–32.

Love, Glen A., *Practical Ecocriticism: Literature, Biology, and the Environment* (Charlottesville: University Press of Virginia, 2003).

Lovecraft, H.P., 'Supernatural Horror in Literature,' *The H.P. Lovecraft Archive* (2009b [1927]) <http://www.hplovecraft.com/writings/texts/essays/shil.asp> (accessed January 2009).

Lovelock, James, *The Ages of Gaia: A Biography of Our Living Earth* (Oxford: Oxford University Press, 1995).

—, *Gaia: A New Look at Life on Earth* (Oxford: Oxford University Press, 1987 [1979]).

—, *The Revenge of Gaia: Why the Earth Is Fighting Back – And How We Can Still Save Humanity* (London: Penguin, 2006).

McArthur, Dan, 'Ethics & Interplanetary Exploration,' *Philosophy Now: A Magazine of Ideas*, 34 (2001/2002), 11–13.

McDowell, Michael J., 'The Bakhtinian Road to Ecological Insight,' in Cheryll Glotfelty and Harold Fromm (eds), *The Ecocriticism Reader: Landmarks in Literary Ecology* (Athens: University of Georgia Press, 1996), pp. 371–91.

McKay, Christopher P., 'Does Mars Have Rights? An Approach to the Environmental Ethics of Planetary Engineering,' in Don MacNiven (ed.), *Moral Expertise: Studies in Practical and Professional Ethics* (London: Routledge, 1990), pp. 184–97.

—, 'On Terraforming Mars,' *Extrapolation*, 23.4 (1982), 309–14.

McNeill, J.R., *Something New Under the Sun: An Environmental History of the Twentieth Century* (London: Penguin, 2001).

McVeigh, Kevin, 'Red Prophet, Green Man, Blue Adept: Kim Stanley Robinson in Conversation With Kevin McVeigh,' *Vector*, 189 (1995), 3–9.

Manning, Laurence, *The How and Why of Better Gardening* (New York: Van Nostrand, 1951).

Marx, Karl and Friedrich Engels, *The German Ideology* (London: Lawrence & Wishart, 1965).

Marx, Leo, *The Machine in the Garden: Technology and the Pastoral Ideal in America* (New York: Oxford University Press, 1964).

Mathisen, Werner Christie, 'The Underestimation of Politics in Green Utopias: The Description of Politics in Huxley's *Island*, Le Guin's *The Dispossessed*, and Callenbach's *Ecotopia*,' *Utopian Studies*, 12.1 (2001), 56–78.

Matthews, Patricia M., 'Aesthetic Appreciation of Art and Nature,' *British Journal of Aesthetics*, 41.4 (2001), 395–410.

Milburn, Colin, 'Greener on the Other Side: Science Fiction and the Problem of Green Nanotechnology,' *Configurations*, 20.1/2 (2012), 53–87.

Morrissey, Thomas J., 'Ready or Not, Here We Come: Metaphors of the Martian Megatext from Wells to Robinson,' *Journal of the Fantastic in the Arts*, 10.4 (2000), 372–94.

Moylan, Tom, *Demand the Impossible: Science Fiction and Utopian Imagination* (London: Methuen, 1986).

'Mukhtar, n.,' *Online Oxford English Dictionary* (2015) <http://www.oed.com/view/Entry/123395?redirectedFrom=Mukhtar#eid> (accessed February 2015).

Murphy, Patrick D., 'Dialoguing With Bakhtin Over Our Ethical Responsibility to Anothers,' in Axel Goodbody and Kate Rigby (eds), *Ecocritical Theory* (Charlottesville: University of Virginia Press, 2011), pp. 155–67.

—, 'An Ecological Feminist Revisioning of the Masculinist Sublime,' *Revista Canaria de Estudios Ingleses*, 64 (2012), 79–94.

—, 'Environmentalism,' in Mark Bould, Andrew M. Butler, Adam Roberts, and Sherryl Vint (eds), *The Routledge Companion to Science Fiction* (London: Routledge, 2009), pp. 374–81.

—, 'The Non-alibi of Alien Scapes: Sf and Ecocriticism,' in Karla Armbruster and Kathleen R. Wallace (eds), *Beyond Nature Writing: Expanding the Boundaries of Ecocriticism* (Charlottesville: University of Virginia Press, 2001), pp. 263–78.

Nudelman, Rafail, 'Labyrinth, Double and Mask in the Science Fiction of Stanislaw Lem,' in Patrick Parrinder (ed.), *Learning From Other Worlds: Estrangement, Cognition and the Politics of Science Fiction and Utopia* (Liverpool: Liverpool University Press, 2000), pp. 178–92.

Oberg, James, *New Earths: Transforming Other Planets for Humanity* (Harrisburg, PA: Stackpole Books, 1981).

Otto, Eric, *Green Speculations: Science Fiction and Transformative Environmentalism* (Columbus: Ohio State University, 2012).

—, 'Kim Stanley Robinson's *Mars* Trilogy and the Leopoldian Land Ethic,' *Utopian Studies*, 14.2 (2003), 118–35.

Pak, Chris, '"A Creature Alive but Tranced and Obscurely Yearning to Wake": Gaian Anticipations and Terraforming in the European Science Fiction of H.G. Wells, Olaf Stapledon and Stanislaw Lem,' *Revista Hélice: Reflexiones Críticas Sobre Ficción Especulativa*, 15 (2012a), 12–19.

—, '"All Energy Is Borrowed" – Terraforming: A Master Motif for Physical and Cultural Re(Up)cycling in Kim Stanley Robinson's *Mars* Trilogy,' *Green Letters*, 18 (2014a), 1–13 <http://www.tandfonline.com/doi/abs/10.1080/14688417.2014.890527> (accessed March 2016).

—, 'Ecocriticism and Terraforming: Building Critical Spaces,' *Forum: University of Edinburgh Postgraduate Journal of Culture & the Arts*, 10 (2010) <http://www.forumjournal.org/site/issue/10/chris-pak> (accessed March 2016).

—, 'Environmental Philosophy and Space in Literature: Landscaping and the Chronotope,' *Brief: Online Journal of Snippets – American Literature*, 1 (2011), 5–6 <http://www.brief.umcs.eu/american-literature/brief-1_american-literature.pdf> (accessed March 2016).

—, '"A Fantastic Reflex of Itself, An Echo, A Symbol, A Myth, A Crazy Dream": Terraforming as Landscaping Nature's Otherness in H.G. Wells's *The Shape of Things to Come* and Olaf Stapledon's *Last and First Men* and *Star Maker*,' *Foundation: The International Review of Science Fiction*, 111.40 (2012b), 14–31.

—, 'An Interview With Frederick Turner,' *SFRA Review*, 307/308 (2014b), 5–12 <http://sfra.org/sfra-review/307-308.pdf>.

—, 'Terraforming and Proto-Gaian Narratives in American Pulp Sf of the 1930s–1940s,' *Eaton Journal of Archival Research*, 1 (2013), 38–55.

Parrinder, Patrick, 'Science Fiction as Truncated Epic,' in George E. Slusser, George R. Guffey, and Mark Rose (eds), *Bridges to Science Fiction* (Carbondale: Southern Illinois University State Press, 1980), pp. 91–106.

Plumwood, Val, *Feminism and the Mastery of Nature* (London: Routledge, 1993).

Prettyman, Gib, 'Daoism, Ecology, and World Reduction in Le Guin's Utopian Fictions,' in Gerry Canavan and Kim Stanley Robinson (eds), *Green Planets: Ecology and Science Fiction* (Middletown, CT: Wesleyan University Press, 2014), pp. 56–76.

Prucher, Jeffrey, 'Terraform,' in *Brave New Words: The Oxford Dictionary of Science Fiction* (Oxford: Oxford University Press, 2007), p. 235.

Rampton, James, 'James Cameron: King of All He Surveys,' *The Independent* (2009) <http://www.independent.co.uk/arts-entertainment/films/features/james-cameron-king-of-all-he-surveys-429268.html> (accessed February 2015).

Rasula, Jed, *This Compost: Ecological Imperatives in American Poetry* (London: University of Georgia Press, 2002).

Rieder, John, *Colonialism and the Emergence of Science Fiction* (Middletown, CT: Wesleyan University Press, 2008).

Rolston III, Holmes, 'Does Aesthetic Appreciation of Landscapes Need to Be Science-Based?' *British Journal of Aesthetics*, 35.4 (1995), 374–86.

Rueckert, William, 'Literature and Ecology: An Experiment in Ecocriticism,' in Cheryll Glotfelty and Harold Fromm (eds), *The Ecocriticism Reader: Landmarks in Literary Ecology* (Athens: University of Georgia Press, 1996), pp. 105–23.

Sagan, Carl, 'Planetary Engineering on Mars,' *Icarus*, 20 (1973), 513–14.

Said, Edward, *Orientalism* (London: Penguin, 2003).

Saito, Yuriko, 'The Aesthetics of Unscenic Nature,' *Journal of Aesthetics and Art Criticism*, 56.2 (1998), 101–11.

Sawyer, Andy, 'Ursula Le Guin and the Pastoral Mode,' *Extrapolation*, 47.3 (2006), 396–426.

Sparrow, Robert, 'The Ethics of Terraforming,' *Environmental Ethics*, 21.3 (1999), 227–45.

Stableford, Brian, 'Ecology and Dystopia,' in Gregory Claeys (ed.), *The Cambridge Companion to Utopian Literature* (Cambridge: Cambridge University Press, 2010), pp. 259–81.

—, *Scientific Romance in Britain 1890–1950* (London: Fourth Estate, 1985).

Stratton, Susan, 'The Messiah and the Greens: The Shape of Environmental Action in *Dune* and *Pacific Edge*,' *Extrapolation*, 42.4 (2001), 303–16.

—, 'Theory and Beyond: Ecocriticism,' *SFRA Review*, 249 (2000), 2–6.

Suvin, Darko, 'Of Starship Troopers and Refuseniks: War and Militarism in U.S. Science Fiction, Part 1 (1945–1974: Fordism),' in Donald M. Hassler and Clyde Wilcox (eds), *New Boundaries in Political Science Fiction* (Columbia: University of South Carolina Press, 2008), pp. 115–44.

—, 'Parables of De-alienation: Le Guin's Widdershins Dance,' *Science Fiction Studies*, 2.3 (1975), 265–74.

'Terraforming, n.,' *Online Oxford English Dictionary* (2015) <http://www.oed.com/view/Entry/240163?redirectedFrom=Terraform#eid12903891> (accessed February 2015).

Thomas, William Leroy, *Man's Role in Changing the Face of the Earth* (Chicago: University of Chicago Press, 1971 [1956]).

Turner, Frederick, *Epic: Form, Content, and History* (New Brunswick: Transaction Publishers, 2012).

—, 'The Humble Bee: Restoration as Natural Reproduction,' *Restoration and Management Notes*, 5.1 (1987), 15–17.

Turney, Jon, *Lovelock and Gaia* (New York: Columbia University Press, 2003).

Vassiliev, M. and S. Gouschev, *Life in the Twenty-First Century*, trans. by H.E. Crowcroft and R.J. Wason (London: Penguin, 1961).

Vera, Hernán and Andrew M. Gordon, 'Dances with Aliens: James Cameron's *Avatar* Movie and White "Saviours" (Updated),' *Racism Review* (2009) <http://www.racismreview.com/blog/2009/12/30/dances-with-aliens-james-camerons-avatar-movie-and-white-saviors/> (accessed February 2015).

Voller, Jack G., 'Universal Mindscapes: The Gaia Hypothesis in Science Fiction,' in George E. Slusser and Eric S. Rabkin (eds), *Mindscapes: The Geographies of Imagined Worlds* (Carbondale: Southern Illinois University State Press, 1989), pp. 136–54.

Westfahl, Gary, 'All Energy Is Borrowed: A Review of *Avatar*,' *Locus Online* (2009) <http://www.locusmag.com/Reviews/2009/12/all-energy-is-borrowed-review-of-avatar.html> (accessed December 2010).

Wiener, Nobert, *Cybernetics: Or Control and Communication in the Animal and the Machine* (Cambridge, MA: MIT Press, 1961 [1948]).

—, *The Human Use of Human Beings* (New York: Da Capo Press, 1988).

Williams, Raymond, *The Country and the City* (Nottingham: Spokesman Books, 2011).

Wolfe, Gary K., *The Known and the Unknown: The Icons of Science Fiction* (Kent: Kent State University, 1979).

Yanarella, Ernest J., *The Cross, the Plow and the Skyline: Contemporary Science Fiction and the Ecological Imagination* (Florida: Brown Walker Press, 2001).

Index

Printed and bound by CPI Group (UK) Ltd, Croydon, CR0 4YY

07/07/2026

14916200-0005